Friction &
Flow Stress
in Forming &
Cutting

INNOVATIVE TECHNOLOGY SERIES
INFORMATION SYSTEMS AND NETWORKS

Friction & Flow Stress in Forming & Cutting

edited by
**Philippe Boisse, Taylan Altan
& Kees van Luttervelt**

First published in 2001 by Hermes Science Publications, Paris
First published in Great Britain and the United States in 2003 by Kogan Page Science,
an imprint Kogan Page Limited
Derived from *International Journal of Forming Processes,* Volume 4, No. 1-2.

First South Asian Edition 2007

Kogan Page Limited
120 Pentonville Road
London N1 9JN
UK
www.koganpagescience.com

Kogan Page India
4737/23 Ansari Road
New Delhi- 110002

© Hermes Science Publishing Limited

© Kogan Page Limited

ISBN 1-9039-9641-4

British Library Cataloguing-in-Publication Data

A CIP record for this book is available from the British Library.

Library of Congress Cataloging-in-Publication Data

Friction and flow stress in forming and cutting / edited by Philippe
Boisse, Taylan Altan and Kees van Luttervelt.
    p. cm.
ISBN 1-9039-9641-4
    1. Metal-work. 2. Friction I. Boisse, Philippe. II. Altan, Taylan.
    III. Luttervelt, Kees van.
TS205.F73 2003
671--dc21

                                          2003003758

Printed in Brijbasi Art Press Ltd., I-72, Sector-9, Noida, U.P. India.

Contents

Foreword

This publication is based on selected papers presented at the International Workshop on Friction and Flow Stress in Cutting and Forming held at the Ecole Nationale Supérieure d'Arts et Métiers (ENSAM) in Paris, France, in January 2000.

During last decade significant developments have taken place in the application of computational mechanics to cutting and forming operations. Both types of operations have in common that very high flow stresses occur at extreme conditions of strain, strain rate, temperature and temperature rate and, that interfaces between tool and workpiece of chip, severe conditions of friction and wear are present which significantly affect the phenomena taking place.

Material behaviour, friction and wear and related phenomena can be studied and the actual values of the relevant flow stress, friction and wear models can be obtained in actual manufacturing operations or in simplified test set ups, like the Hopkinson bar test or like some friction test rig. Nearly none of those tests test the material behaviour or the relevant quantities in actual manufacturing operations and need various assumptions. Usually, the results of such experiments are represented in the form of mathematical relations, which can be used in further computations. Recently doubts have arisen about the value of those models and relations. One aspect is the extrapolation of the experimental data obtained under often simplified conditions in the test set up to the actual conditions in industrial operations. A typical example is that the actual conditions at the tool-chip interface in high speed machining are far away from those in the Hopkinson bar test.

In this publication is presented and discussed the state of art and new developments dealing with flow stress, friction in mechanical processing such as cutting and forming of metals and other materials. Emphasis is put on:

– studying and testing of friction and flow stress in actual industrial processes,

– modelling of friction and flow stress and related phenomena or,

– testing under simulated conditions in order to obtain the data required to model friction and flow stress and related phenomena in actual industrial processes,

– obtaining more detailed information concerning fundamental aspects of recent developments in those processes like the use of coated tools, high speed machining, unusual workmaterials, micro-processing,

– new possibilities to reduce friction and wear.

Industry badly needs more reliable information on flow stress, friction and wear in industrial processes since those phenomena are largely not well understood, unpredictable and cause lots of nuisances like high and unpredictable values of processing forces, temperatures, tool's life, poor precision and surface condition of workpieces.

M. Touratier
C.A. van Luttervelt

Chapter 1

How to Understand Friction and Wear in Mechanical Working Processes

D.A. Taminiau and J.H. Dautzenberg
Dept of Mechanical Engineering, Eindhoven University of Technology, The Netherlands

1. Introduction

In this contribution, an overview will be presented for an understanding of friction and wear using plasticity theory and chemical thermodynamics. It is not worked out in detail, but should, it is hoped, give more than enough proof that this idea is a fruitful line of research.

In a number of mechanical processes like rough and fine cutting, abrading, scraping, punching and dry sliding tests, it has been shown for a number of different metals like copper, steel and aluminum that dry sliding friction is caused by plastic deformation of one of both of the components [DAU 89]. If the process is restricted to plastic deformation only the flow stress or the hardness of the tool material - which has a constant relation to the flow stress - must be higher than that of the workpiece material at the process temperature. When measuring the hardness of the tool material it is important to determine it at the process temperature and not at normal room temperature. Also, it is important to correct the hardness of the workpiece material for temperature, strain rate and the strain path.

To improve tool life for forming operations or to increase cutting speed for improving the economy of the mechanical working processes there is a strong need to look for tool materials of greater hardness at high temperatures. In mechanical working operations most of the tools fail not by tool breakage caused by for instance fatigue but by a continuous rubbing away of the tool material. This article focuses on this wear phenomenon.

The strong demand for longer tool life has led to the use of simple ceramics, composed ceramics and recently to tools with thin layer coatings of complex composition. However, if the hardness under test conditions of the tool is greater than that of the workpiece material, the wear of the tool has to be zero on the basis of mechanics. In practice, though, wear is found. This can only occur if the hardness of the tool material is lowered. This is possible if the composition of the tool material or the coating on the tool changes its chemical composition. It will be shown that this happens by diffusion reactions. As a consequence, the chemical composition of the tool changes and the accompanying hardness of the tool decreases. At a given time the hardness of the tool is lower than that of the workpiece material. The tool material is rubbed off and the tool is considered to be worn. In the following sections it will be proved step by step that this idea holds. Finally, it will be shown that this can be used to understand why some tool/workpiece combinations are unsuitable.

2. Hardness and temperature

Figure 1 shows the hardness at room temperature of binary ceramics, which can be used as a tool material [BHU 91]. From this figure it can be concluded that not only hardness determines the applicability as a tool material, since a number of these ceramics are not used as a tool material in spite of their high hardness. As mentioned earlier, the hardness at room temperature is not important, but the hardness at the process temperature of the mechanical working processes is. In Figure 2 is given the hardness [WES 67-SAN 66] of a number of possible tool materials as a function of temperature. Clearly, in the case of Si_3N_4 the hardness has a marginal dependence on temperature. This tool material can be useful at high temperature, while at low temperatures it is much worse than a number of others. Similar results are found for WC in Figure 3 [WES 67]. In this figure it can be seen that a number of carbides cannot be used at high temperature.

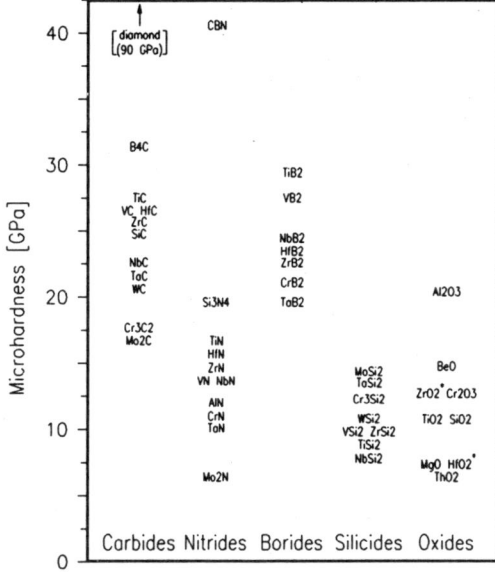

Figure 1. *The hardness of the most important binary ceramics at room temperature*

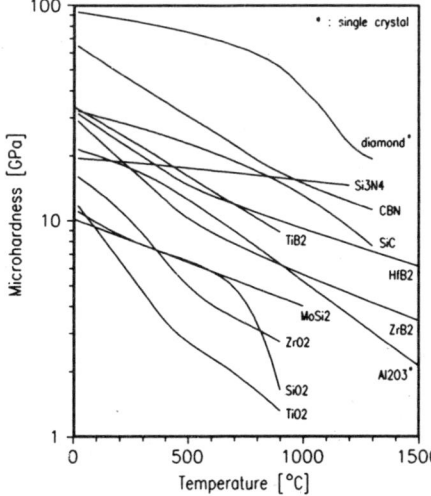

Figure 2. *The hardness of some toll materials as a function of temperature*

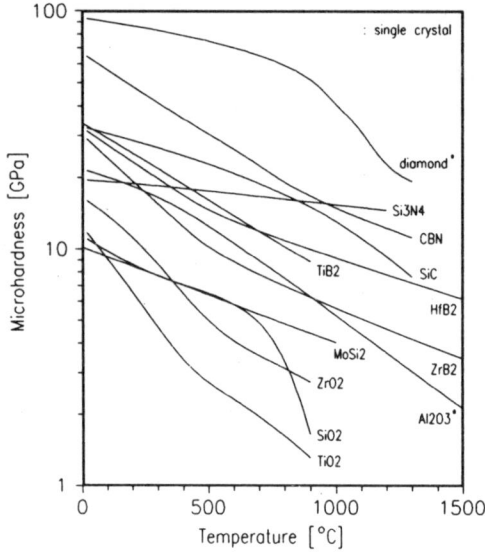

Figure 3. *The hardness of a number of carbides as a function of temperature*

If hardness were the only parameter for wear, diamond would be the best tool material. However, it is known to be unsuitable for ferro-metals.

Next, the flow stress of the workpiece material will be discussed. Figure 4 gives the flow stress behaviour of steel 1045 at different temperatures and at low and high strain rates [JAS 99]. The influence of both variables is clearly shown. From tests on a number of workpiece materials it is clear that the mechanical material behaviour is still theoretically unpredictable. The only way to find the actual flow stress behaviour is to measure it under the same process conditions as the mechanical working process.

To obtain an idea of process temperatures, two examples will be given. Cutting steel AISI1045 with a carbide tool at 4 m/s, a feed of 0,2 mm and a width of cut of 4 mm, gives a mean contact temperature of approximately 750°C and a contact time of approximately 0.5 ms [JAS 98]. Such temperatures of 700°C or more are possible in the shear zone by punching a 1mm strip of 13% Cr low carbon construction steel at 50 mm/s punch speed [BRO 99].

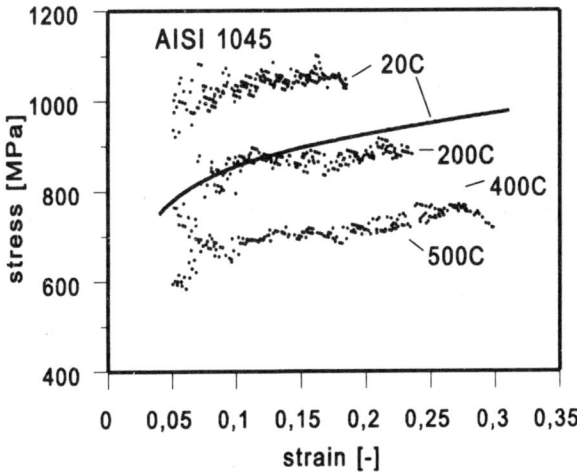

Figure 4. *Flow stress behaviour of steel AISI 10455 at different temperatures and strain rates (dots : $\dot{\varepsilon} = 7.5 \times 10^3 \ s^{-1}$, solid line : $\dot{\varepsilon} = 0.006 \ s^{-1}$)*

3. Diffusion

Besides temperature, the diffusion of tool material in the workpiece material or vice versa is also important for the hardness. In Figure 5 an example is given of diffusion of tungsten in a steel chip when machining AISI 1045 steel with a cemented carbide tool. The cutting conditions are as follows: cutting speed 4m/s, feed 0,2 mm, and width 4mm. The cutting tool is a Widia insert TPGN110204 THM-F without coating. The diffusion pattern is measured by Rutherford backscattering spectroscopy with 4 MeV He+ ions. Figure 5 shows the concentration of tungsten that has diffused from the tool into the chip as a function of the distance to the contact surface of chip and tool. The solid line in this figure is an erf-function as expected for a diffusion process. Although the cutting speed is given by the tool manufacturer, it results in a very clear diffusion pattern. The same can be found at lower cutting speeds but less markedly so.

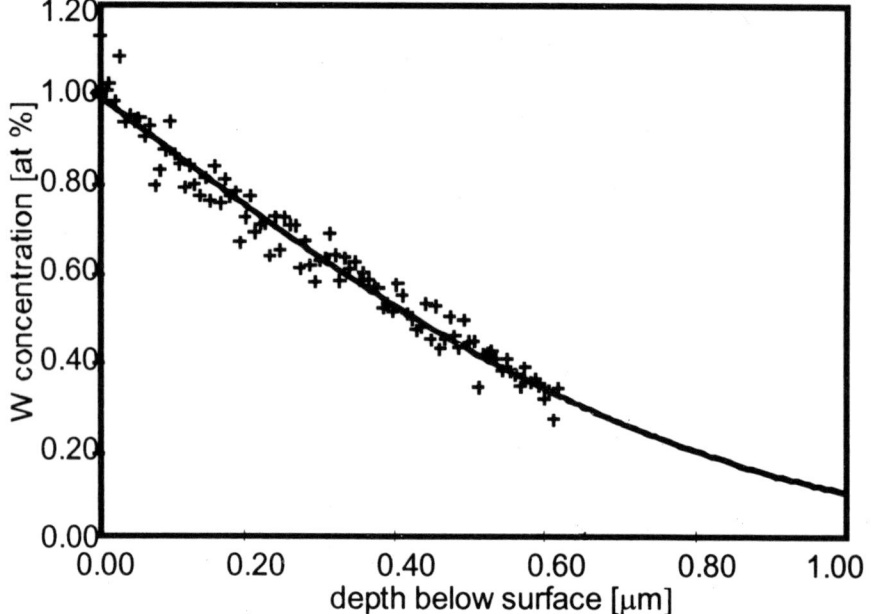

Figure 5. *Tungsten concentration as a function of the distance to the contact surface of tool and chip. Cutting speed 4 m/s; feed 0.2 mm. Tool: Widia TPGN 110204 THM-F; workpiece material: steel AISI 1045*

The same is true for all tool materials that fail in dry sliding friction by continuous abrading wear.

For a better understanding of the possibility of diffusion [KRA 80, KRA 85, SUH 86] we consider a very simple tool/workpiece combination A/B. The diffusion of A in B is driven by the free enthalpy ($= \Delta G_m$) of mixing given by:

$$\Delta G_m = \Delta H_m - T \Delta S_m \qquad\qquad [1]$$

$$\Delta H_m = H_{AB} - H_A - H_B \qquad\qquad [2]$$

$$-T\Delta S_m = RT (x_A \ln x_A + x_B \ln x_B) \qquad\qquad [3]$$

where:

- ΔH_m = enthalpy of mixing
- ΔS_m = entropy of mixing
- H_i = enthalpy of component i
- x_i = concentration of component i

and: $x_A + x_B = 1$.

The diffusion of A in B or vice versa only takes place if $\Delta G_m < 0$. The most ideal tool/workpiece material combination is if $H_{AB} = 0$ and H_A and H_B are very negative. This means no reaction between tool and workpiece material will take place. The lowest concentration of A is $1/6 \times 10^{-23}$, which means that $x_B = 1$. Completing equation [3] with $T = 1\,000°K$ and expanding it for a gram-atom A gives:

$$T\Delta S_m = -449\,kJ \tag{4}$$

Filling in this result in [1] means that we can calculate ΔG_m if we know ΔH for our tool and workpiece material.

Figures 6 to 10 give the enthalpy or Gibbs Energy [KUB 79, BAR 73] of formation for carbides, nitrides, oxides, silicides and borides. A combination of these figures with equations [4] and [1] means that if the temperature is high enough ΔG_m is always negative. So diffusion will always take place.

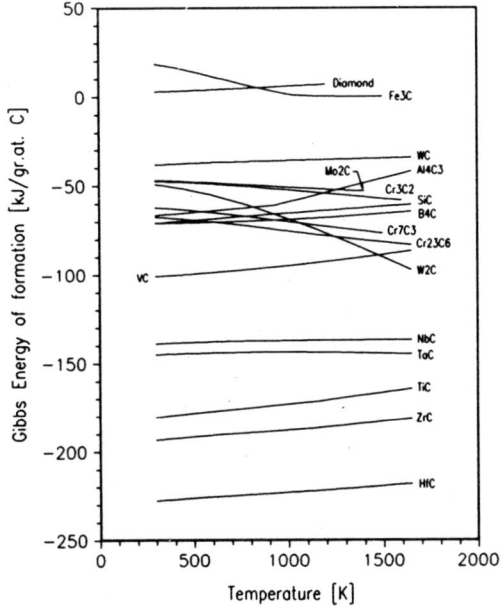

Figure 6. *The enthalpy or Gibbs energy of formation of carbides*

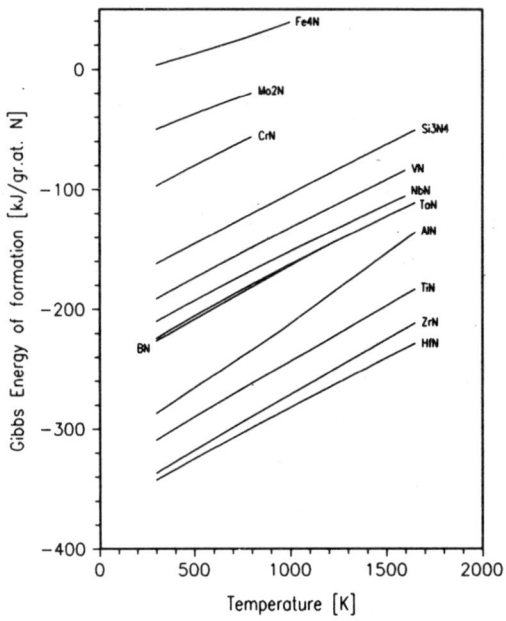

Figure 7. *The enthalpy or Gibbs energy of formation of nitrides*

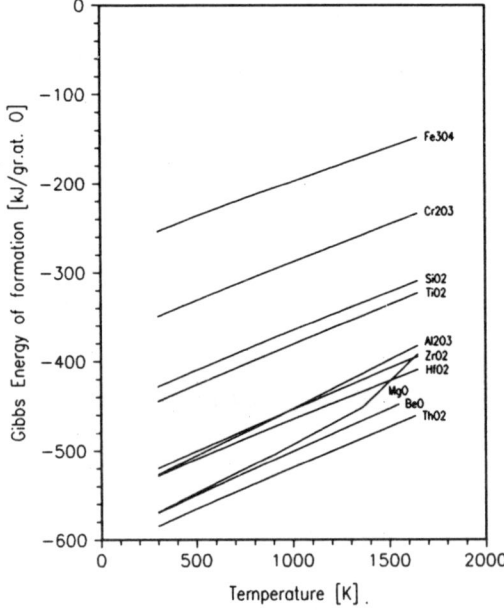

Figure 8. *The enthalpy or Gibbs energy of formation of oxides*

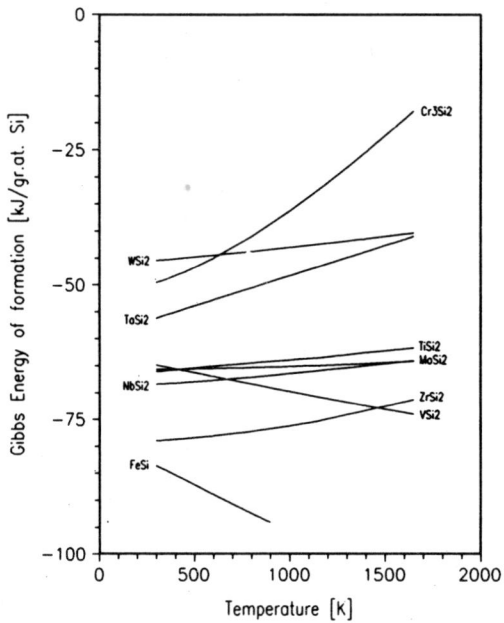

Figure 9. *The enthalpy or Gibbs energy of formation of silicides*

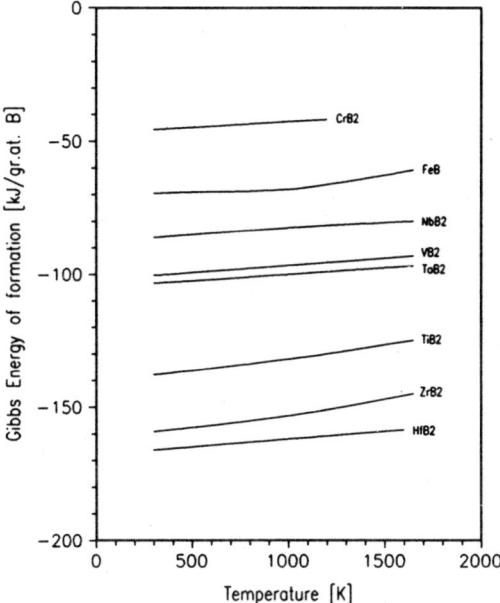

Figure 10. *The enthalpy or Gibbs energy of formation of borides*

If steel is used as a workpiece material, it is important that the tool material has a much more negative enthalpy in comparison with any possible combination of iron with one of the elements of the tool material.

The following examples are all based on iron (steel) as a workpiece material.

In Figure 6 it is found that around 700°C Fe_3C is more stable than diamond. Therefore diamond is completely unsuitable at high temperatures in a mechanical working process in combination with iron or steel. However, at room temperature diamond is more stable and usable with steel.

From Figure 6 we can conclude that HfC, ZrC, TiC, TaC and NbC are more useful as tool materials than WC, regarding the enthalpy of these carbides. In cutting tools for steel TiC and TaC are preferred above WC.

From Figure 9 it is seen that silicides are completely unsuitable as a tool material for ferro-metals.

Figure 8 shows that the oxides are very stable and are suitable for tool materials despite their relatively low hardness.

4. Hardness and chemical composition

In the preceding sections we have seen that diffusion of the tool material in the workpiece material takes place and vice versa. This means that the chemical composition of the tool material changes. Figure 11 shows that with this change in chemical composition a change in hardness takes place [HOL 86]. Combined with the high tool temperature and the low workpiece temperature at the beginning of the mechanical working process, it is clear that a moment comes when the hardness of the workpiece material is higher than the hardness of the tool material and the tool material is worn as a result. In contrast with the previous figures, Figure 11 is made at room temperature. To the authors' knowledge there are no figures available at higher temperatures at this time.

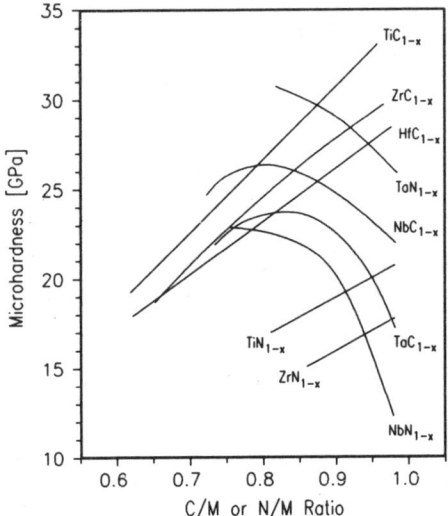

Figure 11. *The microhardness as a function of the chemical composition of some carbides and nitrides*

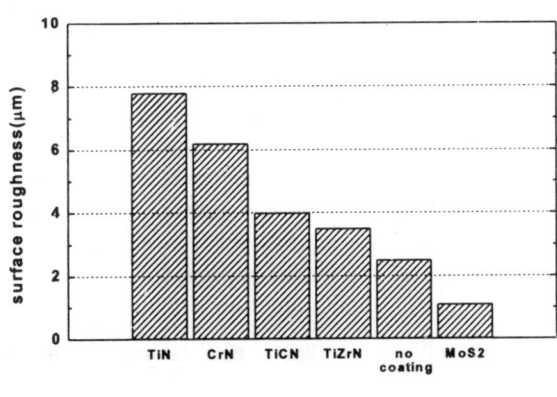

coating onto HSS

Figure 12. *Overall performance improvements in terms of workpiece surface finish using the MoS₂ coating in end milling (φ 16mm) of an Al-Mg alloy*

5. Results

In the preceding text, some general examples have already been given.

Another well-known example is the use of TiN as a tool for aluminum. It is known that it has a very high wear in cutting as well in forming. If we look at Figure 7 this behaviour is obvious. A better tool can be achieved by adding aluminum in TiN to

TiAlN. A much better solution is possible if we replace N by C. In Figure 6 TiC is very negative whereas Al, which is not in this figure, is around zero. This latter assertion is supported by the phenomenon that aluminum can be melted in a graphite crucible.

6. Discussion

It has been shown that with classical physics it is possible to understand friction and wear without using the phenomenon of adhesion. After all, with adhesion theories it is difficult to understand friction and impossible to handle it quantitatively.

The different properties that play a role in making a good tool material make it difficult to design an optimum tool material. This is further complicated by the role of temperature in chemical processes like diffusion, which is very dominant. An increase of temperature of 10 degrees increases the reaction speed by a factor 2 or 3. In mechanical working processes the temperature is strongly influenced by friction. At present the knowledge for predicting the process temperature more accurately than a few degrees is not available. So it has been impossible to calculate wear rates up to now.

To improve tool materials, especially in cutting, people have relied completely on finding ceramic coatings made of more than two elements in order to have a high hardness at high temperatures. Now however attention is focussing elsewhere. It has shifted to tool materials that have a low friction coefficient in dry cutting. This lowers the process temperature and although these materials have a modest hardness, the combination gives a longer tool life. An example of this line of reasoning is the use of MoS_2 [TEE 97, REC 93].

By PVD magnetron sputtering of MoS_2 on a tool it can be shown (Figure 12) that the tool life is enhanced and, moreover, as a consequence of the lower friction the accuracy and the roughness of the products is improved [ENG 83]. In the near future these low friction coatings will be further improved if combined with high hardness materials.

Another method is to alloy the workpiece material with a component that lowers the friction [ENG 83].

A well-known method is fine cutting of nickel with diamond. Cutting pure nickel with diamond is a catastrophe; cutting electrolytic nickel including phosphorus (low friction) with a diamond tool, however, gives excellent results.

If friction is plastic deforming, the high deformations that are necessary in the contact zone might be found to be striking. In [DAU 89], however, it was shown that these extreme deformations do indeed take place in the contact zone. This is because of the high compressive (hydrostatic) stresses in this zone, together with the relatively high temperature of the friction processes that the material can dynamically recrystallise and reobtains its deformation ability. Also it is shown that the fracture strain is strongly dependent on the compressive (hydrostatic) stress [DAU 99]. This makes it possible to understand, without adhesion theory, that a material can fail below the contact surface through which a layer breaks out.

Since the process temperature is sensitive to the thermal properties of tool and workpiece material, it is obvious that thermal properties have a distinct influence on the diffusion processes and thus on wear.

7. Conclusion

Using plasticity theory and chemical thermodynamics it is possible to understand wear in a qualitative way.

Use of thermodynamics makes it easier to select tool materials for workpiece materials.

More attention has to be paid to combinations of materials of low friction and high hardness as tool materials.

Taking into account the influence of hydrostatic stresses on the fracture strain of metals, it is possible to understand the phenomenon of layer breakout during dry sliding friction.

Acknowledgements

The authors are indebted to the European Community for financing this research within a BRITE/EURAM program (project No BE-7239) and also to the Minister of Economic Affairs of The Netherlands for financing this research within a IOP-Metalen project.

Regarding the diffusion measurements the authors are indebted to Dr. L.J. van Ijzendoorn of the Cyclotron Laboratory of Eindhoven University of Technology.

8. References

[BAR 73] BARIN I., KNACKE O., *Thermochemical properties of inorganic substances*, Springer Verlag, 1973.

[BHU 91] BHUSHAN B., GUPTA B.K., *Handbook of tribology; materials, coatings and surface treatments*, McGraw-Hill, 1991.

[BRO 99] BROKKEN D., Numerical modeling of the metal blanking process, Ph.D.-thesis, Eindhoven University of Technology, The Netherlands, 1999.

[BSE 74] BSENKO L., LUNDSTRÖM T., "The high temperature hardness of ZrB_2 and HfB_2", *J. Less. Comm. Met.*, vol. 34, p. 273-278, 1974.

[DAU 89] DAUTZENBERG J.H., van DIJCK J.A.B., KALS J.A.G., "Metal structures by friction in mechanical working processes", *Annals of the CIRP*, 38/1, p. 567-570, 1989.

[DAU 99] DAUTZENBERG J.H., JASPERS S.P.F.C., TAMINIAU D.A., "The workpiece material in machining", *Int. J. Adv. Manuf. Technol.*, 1999.

[ENG 83] ENGSTRÖM U., "Machinability of sintered steels", *Powder Metallurgy*, vol. 26, p. 137-144, 1983.

[HOL 86] HOLLECK H., "Material selection for hard coatings", *J. Vac. Sci. Technol. A*, vol. 4, p. 2661-2669, 1986.

[JAS 98] JASPERS S.P.F.C., DAUTZENBERG J.H., TAMINIAU D.A., "Temperature measurement in orthogonal metal cutting", *Int. J. Adv. Manuf. Technol.*, vol. 14, p. 7-12, 1998.

[JAS 99] JASPERS S.P.F.C., Metal cutting mechanics and material behavior, Ph.D.-thesis, Eindhoven University of Technology, The Netherlands, 1999.

[KRA 80] KRAMER B.M., SUH N.P., "Tool wear by solution: a quantitative approach", *J. Eng. Ind.*, vol. 102, p. 303-309, 1980.

[KRA 85] KRAMER B.M., JUDD P.K., "Computational design of wear coatings", *J. Vac. Sci. Technol. A*, vol. 3, p. 2439-2444, 1985.

[KUB 79] KUBASCHEWSKI O., ALCOCK C.B., *Metallurgical thermochemistry*, Pergamon Press, 1979.

[LOL 63] LOLADZE T.N., BOKUCHAVA G.V., DAVIDOVA G.E., "Temperature dependencies of the microhardness of common abrasive materials in the range of 20 to 1 300 °C" in Westbrook, J.H., Conrad, H., (eds.), *The science of hardness testing and its research applications*, American Society for Metals, 1963.

[NII 84] NIIHARA K., HIRAI T., "Hot hardness of CVD Si_3N_4 to 1 500 °C", *Powder Metall. Int.*, vol. 16, p. 223-226, 1984.

[REC 93] RECHBERGER J., DUBACH R., "Soft PVD coatings-a new coating family for high performance cutting tools", *Mat. wiss. u. Werkstofftechn.*, vol. 24, p. 268-270, 1993.

[SAN 66] SANDERS W.A., PROBST H.B., "Hardness of five borides at 1 625 °C", *J. Am. Ceram. Soc.*, vol. 49, p. 231-232, 1966.

[SUH 86] SUH N.P., *Tribophysics*, Prentice-Hall, 1986.

[TEE 97] TEER D.G., HAMPSHIRE J., FOX V., BELLINDO-GONZALES V., "Tribological properties of MoS_2 metal composite coatings deposited by close-field magnetron sputtering", *Surf. Coat. Technol.*, vol. 94-95, p. 572-577, 1997.

[WES 67] WESTBROOK J.H., STOVER E.R., "Carbides for high-temperature applications" in Campbell I.E., Sherwood E.M., (eds.) *High temperature materials and technology*, Wiley & Sons, 1967.

[WES 66] WESTBROOK J.H., "The temperature dependence of hardness of some common oxides", *Rev. Hautes Tempér et Refract.*, vol. 3, p. 47-57, 1966.

Chapter 2

Friction During Flat Rolling of Metals

John G. Lenard

University of Waterloo, Ontario, Canada

1. Introduction

During commercial flat rolling of steel and aluminium strips control of the coefficient of friction at the roll/work piece contact is achieved by careful choice of the lubricant. The choice also affects the productivity and the surface quality of the rolled strips. Improving these is the prime objective of the producers. At the present time oil-in-water emulsions are used when cold rolling steel or hot rolling aluminium. Neat oils are used mostly when cold rolling aluminium. In most applications the twin objectives are cooling of the surfaces and the provision of sufficient amount of oils in the roll/strip contact zone.

Control of the coefficient of friction presupposes knowledge of its magnitude. While there is a reasonable understanding of the frictional mechanisms at the roll/strip interface during flat rolling, the actual magnitude of the coefficient of friction there is still largely a matter of conjecture. In this context it is appropriate to quote Roberts [ROB 97]:

"Of all the variables associated with rolling, none is more important than friction in the roll bite. Friction in rolling, as in many other mechanical processes, can be a best friend or a mortal enemy, and its control within an optimum range for each process is essential."

The transfer of thermal and mechanical energy at the roll/metal interface is responsible for the quality of the resulting surfaces. In cold rolling, inappropriate magnitudes of friction forces cause unacceptable surfaces. In hot rolling surface defects, accelerated roll wear and unsuitable strain distributions, leading to unwanted grain size distributions, will occur.

The tribological system is defined in terms of process and material parameters [SCH 83], including the temperature, speed and the reduction. Properties of the rolls and the rolled material: their strength, hardness, Young's and shear moduli, stored elastic energy and their thermo-physical properties also affect the interactions. The effects of surface parameters, such as the chemical reactivity, the tendency to adsorb molecules from the environment, surface energy and surface roughness need to be understood. The lubricant's viscosity, bulk modulus, chemical composition, additives and lubricant delivery also contribute to product quality. While all of these should be included in analyses, it is understood that selection of the more important parameters may reduce complexity. In the present study the reduction and the speed are of prime importance and these, in addition to the lubricants, are chosen to be the independent variables.

There are two ways to determine μ in the rolling process: direct measurements and inverse analyses. In the former, tension may be applied to the strip, moving the neutral point to the exit, and inferring the magnitude of the coefficient from the roll force and torque [UND 50]. Also, the minimum coefficient of friction may be identified at the reduction when no roll bite occurs. Transducers embedded in the roll [SIE 33; ROO 57; LIM 84; JES 91; JES 00] give the roll pressure, the interfacial shear stress and their ratio, the coefficient of friction. The inverse method can also be used to infer what the coefficient of friction must have been in a pass [LIN 91].

The objective of the present study is to demonstrate a reliable method to determine μ in the flat rolling process – the embedded pin–transducer combination. A mathematical model, able to predict the coefficient of friction in a consistent manner, is then employed to substantiate the experimental data and to use roll force, torque and forward slip values to infer μ during cold rolling of steel and aluminium strips. The results are discussed in terms of the adhesion hypothesis.

2. Experimental determination of the coefficient of friction

2.1. *The rolling mill*

A two-high laboratory rolling mill, driven by a 42 kW, constant torque DC motor and having D2 tool steel rolls of 250 mm diameter and 150 mm length, hardened to Rc = 64, is used in the study. The surface roughness of the work rolls, measured using a portable roughness tester, is Ra = 0.43 µm in the axial direction and 0.12 µm in the circumferential direction. Two force transducers, located under the bearing blocks of the lower work roll measure the roll separating force. Two torque transducers, in the drive spindles, monitor the roll torque. The time difference between the signals of two photodiodes, positioned at the exit 50.68 mm apart, provides the exit velocity of the rolled strips, leading to the forward slip. A shaft encoder monitors the rotational velocity of the roll, allowing use of the actual motor speed under load when calculating the forward slip (schematic diagram in Figure 1).

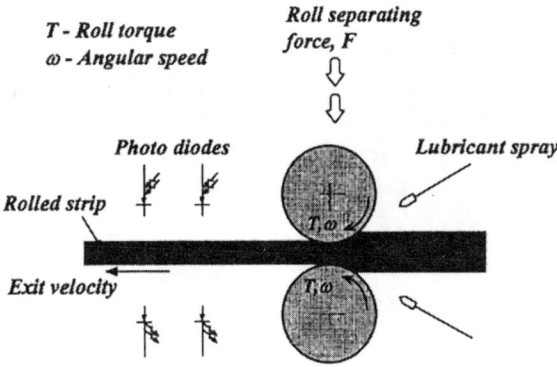

Figure 1. *The schematic diagram of the experimental rolling mill*

2.2. The embedded pin–transducer combination

Originally suggested in [SIE 33] for the rolling process and adapted later [ROO 60; ALS 73], the method has been applied to measure interfacial conditions in several bulk metal forming processes, including forging and extrusion. The method has been used in warm and cold, flat rolling of aluminium strips [KAR 85; LEN 93; LIM 84]. Variations have also been presented [LEN 90; LEN 91; YON 87; YON 89]. A cantilever or cone, fitted with strain gauges, with its tip in the contact zone, and its various refinements were presented in [BAN 72; JES 91;,JES 00]. Detailed information on the distributions of interfacial frictional shear stresses and die pressures may be obtained by these methods, but the experimental set-up and the data acquisition are elaborate and costly. Since the major criticism concerns the possibility of some metal or oxide intruding into the clearance between the pins and their housing [STE 83], it is necessary to substantiate the resulting coefficients of friction by independent means.

In the present study, the roll pressures and the interfacial shear stresses are measured by four pin–transducer combinations, as shown in Figures 2a and 2b [KAR 85]. Figure 2a shows the four transducers, placed in the top work roll. Figure 2b shows the four pins, placed in the segment, which, when put in position, completes the work roll surface. The segment is made of D2 tool steel, hardened to the same magnitude as the work roll. Note that two of the pins are in the radial direction and the two others are placed obliquely, 25° from the radial direction. The tips of the pins are in direct contact with the rolled strip while their flat ends are pressing directly on the force transducers. A detailed description of the apparatus and the analyses necessary to extract the roll pressures, the interfacial shear stresses and hence, the coefficient of friction, have been given earlier [LIM 84; KAR 85].

Figure 2a. *Embedded transducers*   **Figure 2b.** *Measuring pins*

A simple force analysis of the four pins, which includes the signals of the transducers, yields the roll pressures, shear stresses and hence, μ, which varies from entry to exit in the roll gap. There is a certain amount of clearance, of 0.02 mm magnitude, in between the pins and their housing. The analysis includes the friction forces and the pressures exerted by the walls of holes against which the pins may be pressed during a pass. Thus, when contamination, oxides, rolled metal or scales intrude into the clearance, the analysis takes account of their influence as well. Nevertheless, frequent disassembly and cleaning are necessary for continued, reliable testing.

3. The experimental program

The first set of tests is concerned with hot rolling of aluminium strips, lubricated by oil-in-water emulsions. The embedded pin–transducer system is used to determine the coefficient of friction. These values of μ are then used in a mathematical model to calculate the roll force, torque and the forward slip. Since the model is shown to be accurate and consistent in its predictions, it is then used in several other experiments to infer the coefficient of friction.

3.1. *Hot rolling of aluminium strips*

Roll pressure and shear stress distributions have been measured during hot rolling, at 500 °C, of commercially pure aluminium strips, measuring 6.27x50x200 mm, using the embedded pin–transducer method, as described above [HUM 96]. Each sample had a type K thermocouple embedded in its tail end, monitoring the temperature variations during the pass. The emulsion, 2% (v/v) oil and water, was prepared at 60°C and sprayed at the entry of the strips. Typical pressure and shear stress distributions are shown in Figures 3, 4 and 5.

In general, both the roll pressure and the shear stress distributions are similar for the three cases. The distributions of the roll pressure are quite flat after rising rapidly from zero - not shown on the plots due to some uncertainty of the exact entry and exit locations, followed by a rise until the peak is reached. This increase, caused by the strain and strain rate hardening of the plastically flowing metal, is not expected to be affected by the falling temperatures of the rolled strip and the attendant increase in the flow strength since the drop of temperature in the pass, from entry to exit, was not very large, typically less than 10°C. The fall of the surface temperature is also not very pronounced [LEN 93]. In that study temperatures measured by thermocouples were employed as the initial conditions in a finite element analysis of the process. The surface temperatures obtained were only marginally below the temperatures at the strips' centres. The variations of the interfacial shear stresses are also shown in the figures.

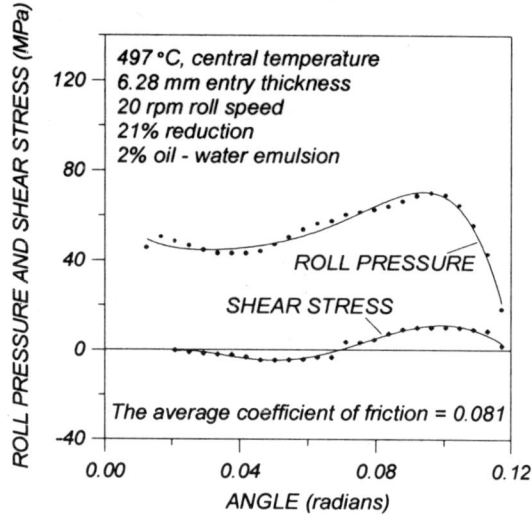

Figure 3. *Roll pressure and shear stress distribution; 21% reduction*

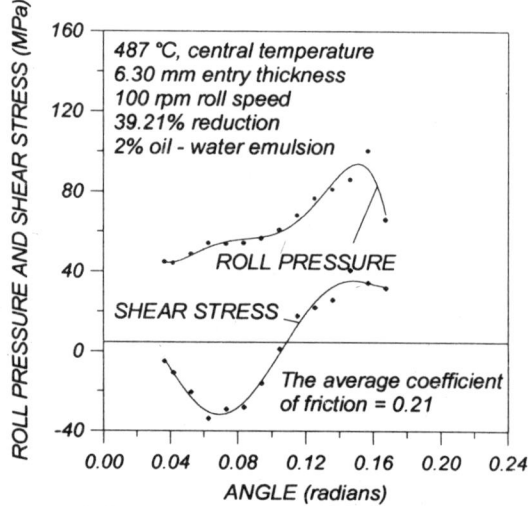

Figure 4. *Roll pressure and shear stress distribution; 39% reduction*

The coefficient of friction is defined as the average of the ratio of the shear stress and the roll pressure. Its variation with the rolling speed and the reduction is shown in Figure 6. The coefficient drops with increasing speed, as found in most instances. It increases when the reduction grows, a phenomenon strongly dependent on the interaction of several parameters. These include the lubricant viscosity and its sensitivity to the pressure and the temperature. The flow strength of the metal and its strain and strain rate hardening also need to be considered as they affect the flattening of the asperities and the attendant growth of the true area of contact. Since the coefficient of friction increases with the loads, the effects of the flattening of the asperities and the growing number of adhesive bonds appear to overwhelm the mechanisms that may cause a drop in its magnitude.

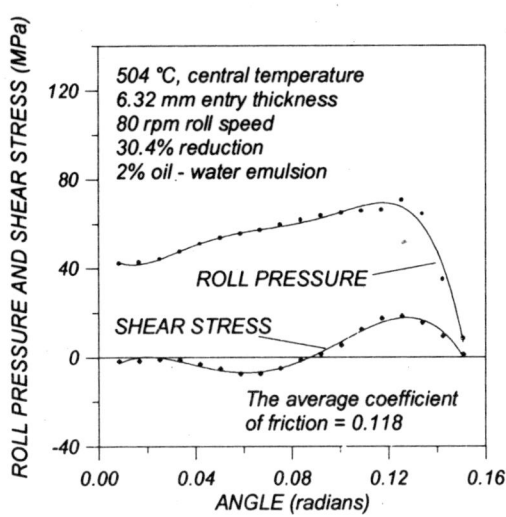

Figure 5. *Roll pressure and shear stress distribution; 30.4% reduction*

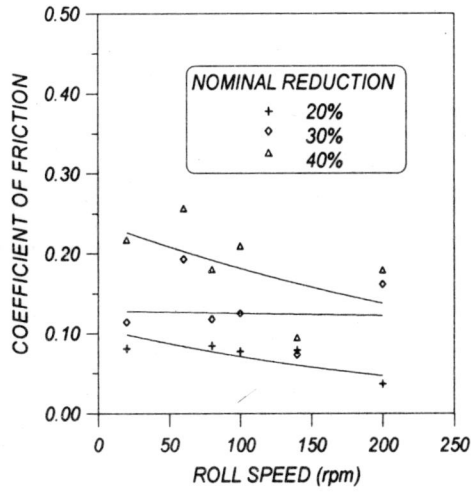

Figure 6. *Coefficient of friction as a function of speed and reduction*

3.2. Substantiation of the measured coefficients of friction

The reliability of the pin–transducer technique has been questioned [STE 83], necessitating proof of its accuracy, which is done in three independent ways. The roll separating forces, measured by force transducers, located under the bearing blocks of the bottom roll, are compared to the integrals over the contact surface of the roll pressure distributions, as indicated by the embedded pin–transducer combinations. The torques, measured by the torque transducers in the spindles of the drive train, are compared to the integrals of the shear forces times the roll radius. The measured coefficients of friction are also used in a predictive model of the process, which calculates the roll separating forces, the roll torques and the forward slip. When the measured and computed forces, torques and the forward slip match, the measured coefficient of friction is accepted as the correct value.

There is experimental evidence that the coefficient of friction does not remain constant in the roll gap, see Figures 3, 4 and 5, above. The model, given in detail in [LEN 97] and described briefly below, allows the use of a coefficient of friction, which has different magnitudes on either side of the neutral point.

As is well known, Orowan's model [ORO 43] uses the friction hill, in which the location of the neutral point is obtained at the intersection of the roll pressure curves, extending from entry and exit. The equations of equilibrium are written, using a functional form:

$$\frac{dp}{d\phi} = f\left(p, 2k, R, h_1, h_2, E, v, \pm\mu p\right) \qquad [1]$$

and these are integrated separately, using the appropriate algebraic signs for the friction terms. The neutral point coincides with the location of the intersection of the two curves. In the present refinement, only one equation of equilibrium, of the form:

$$\frac{dp}{d\phi} = f\left[p, 2k, R, h_1, h_2, E, v, \mu(\phi)\right] \qquad [2]$$

is employed. An assumption for the variation of the coefficient of friction in the roll gap is made, with some guidance from previous experience. This assumption includes the location of the neutral point. The coefficient is taken to be positive between the entry and the neutral point and negative beyond it, changing gradually at the no-slip location. The functional form is

$$\mu = \mu(\phi) \qquad [3]$$

The equation of equilibrium is then integrated, starting with the known initial condition at the entry. Satisfaction of the boundary condition at exit drives the first iterative process. The second iteration is designed to achieve convergence of the predicted roll separating forces. The results are shown in the table below, giving the forward slip, the roll force and the torque in the top spindle.

h_{in}	Red	rpm	μ	S_f, %		Roll force, kN/mm			Roll torque, Nm/mm		
mm	%			test	M.	test	Int	M.	test	Int	M.
6.28	21.0	20	.081	1.11	1.5	.76	.76	.80	3.94	3.89	3.48
6.31	21.6	200	.037	3.15	3.09	.97	.99	.98	4.69	4.65	4.64
6.29	30.7	20	.114	2.57	2.74	1.3	1.21	1.28	8.3	8.26	7.87
6.32	31.8	200	.162	2.52	2.68	1.06	.94	.96	5.36	5.83	5.75
6.30	39.2	100	.210	4.56	4.82	1.46	1.31	1.31	8.88	9.22	9.25

Use of the experimentally determined coefficients resulted in the measured and calculated values which are sufficiently close to conclude that the embedded pin–transducer technique is capable of producing accurate values. The model is also shown to be valid and it may be used with confidence to infer the magnitude of the coefficient. The symbol M in the table refers to the model's predictions.

3.3. Cold rolling of steel strips, using neat oils

Cold rolling of steel strips, lubricated using neat oils, is the next step in the study [MCC 00]. The steel contains 0.05% C and its true stress - true strain curve is $\sigma = 150(1 + 234\varepsilon)^{0.251}$ MPa, obtained in uniaxial tension. The metal's strength is essentially independent of the rate of strain. The initial surface roughness of the samples is approximately Ra = 0.8 μm, in both the rolling and transverse directions.

Six oils are used in the tests, identified as A, B, C, D, E and F. A is a petroleum based oil with sulfurised hydrocarbons, fats and esters. B is identical to A, but with no additives. C is again the same, containing a lubricity contributing ester. D is a sulfurised petroleum based oil of low viscosity. E contains small amounts of esters as additives. Oil F is a synthetic lubricant with no additives. The properties of the lubricants are shown in the table below.

Properties of the lubricants

Lubricant	Kinematic viscosity (mm^2/s)		Density (kg/m^3)
	40°C	100°C	40°C
A	25.15	4.97	869.3
B	19.85	3.96	861.6
C	20.05	4.03	863.0
D	14.30	3.28	886.6
E	5.95	1.88	853.1
F	17.32	3.89	819.4

The roll force, roll torque and the forward slip have been measured and used in the inverse analysis with the model described above, to infer the magnitude of the coefficient of friction. The computed coefficients of friction for each of the lubricants indicate similar trends for all six oils. The results are shown in Figure 7 for the reductions of 15 and 50%. The coefficient of friction is reduced as the speed and the reduction are increased. While the lowest coefficient of friction is obtained with lubricant F and the highest with A and D, the differences among the lubricants are not large. As well, the additives appear to affect friction more than does the viscosity. The coefficients of friction may also be determined by several empirical formulae. The most often used formula is due to Hill [HOF 53].

$$\mu = \frac{\dfrac{P_r}{\bar{\sigma}\sqrt{R'\Delta h}} - 1.08 + 1.02\left(1 - \dfrac{h_{exit}}{h_{entry}}\right)}{1.79\left(1 - \dfrac{h_{exit}}{h_{entry}}\right)\sqrt{\dfrac{R'}{h_{entry}}}} \qquad [4]$$

where P_r is the roll separating force per unit width, $\bar{\sigma}$ stands for the average, plane strain flow strength in the pass and R' designates the radius of the flattened roll, calculated using Hitchcock's relation. Using Hill's formula, see Figure 8, indicates that the more refined model predicts the same trends but significantly lower magnitudes for the coefficient of friction. Also demonstrated is the importance of the model as the results of the calculations depend on it in a very significant manner.

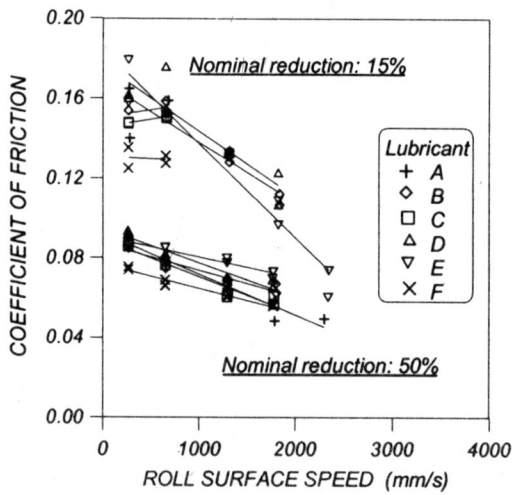

Figure 7. *Coefficient of friction, obtained using a 1D model; cold rolling of steel strips*

As expected from previous tests and as predicted by the adhesion hypothesis, the coefficient of friction decreases as the roll surface speed increases. Further, the coefficient falls as the reduction and hence, the load, increase.

It may be concluded that Hill's relation is useful if comparative results are required, such as in the study just described. The objective was to rate the six lubricants according to their ability to lower the loads on the rolling mill. As long as the trends, predicted by the model, are realistic, the actual magnitudes of the coefficient of friction are of little importance. Several formulas, which may be may be used to determine μ can be found in the technical literature. The trends they predict should be checked before use.

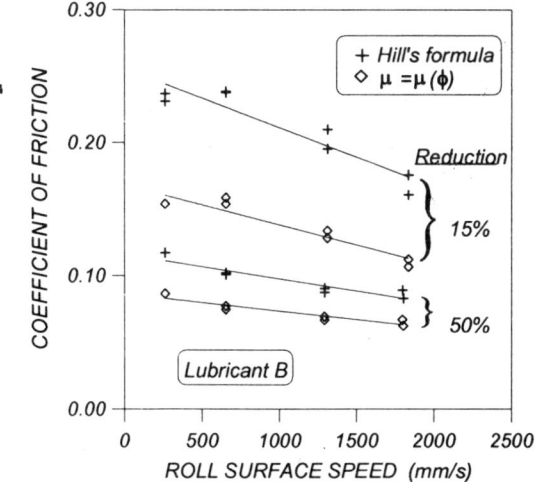

Figure 8. *A comparison of the predicted coefficients by a 1D model and by Hill's formula*

3.4. *Cold rolling of steel strips, comparing neat oils and emulsions*

Low carbon steels are used with the true stress – true strain equation, obtained in uniaxial tension, $\sigma = 174.9(1 + 120.66\varepsilon)^{0.245}$ MPa, in the next set of tests [SHI 00]. Four lubricants are compared in the study, in addition to dry rolling and using water only, for their abilities to affect the loads on the mill, the frictional conditions and the resulting roughness. The lubricants include the SAE 10 and SAE 60 automotive oils, the SAE 10 base oil with 5% oleic acid added and the SAE 10 base oil, emulsified, using water and polyoxyethylene lauryl alcohol as the emulsifier, 4% by volume. Oleic acid was chosen as the boundary additive since it was shown to react to pressure and temperature in a less sensitive manner than several other fatty oils [SCH 87]. The properties of the two base oils are given in the table below.

While the automotive lubricants were not formulated for use in the bulk forming of metals or, specifically for the flat rolling process, their properties – such as their viscosities, densities, pressure and temperature coefficients - are well known, and that is the reason for their choice in this comparative study. The coefficients of friction, at 20 and 160 rpm roll velocity, are shown in Figures 9 and 10. In determining the coefficient, Hill's formula was employed.

Properties of base oils

		kinematic viscosity, mm²/s		density
	Description	40°C	100°C	gm/cm³, 15°C
SAE 10	Paraffinic, refined, dewaxed	29.6	4.9	0.871
SAE 60	Paraffinic, refined, dewaxed	283.7	22.6	0.875

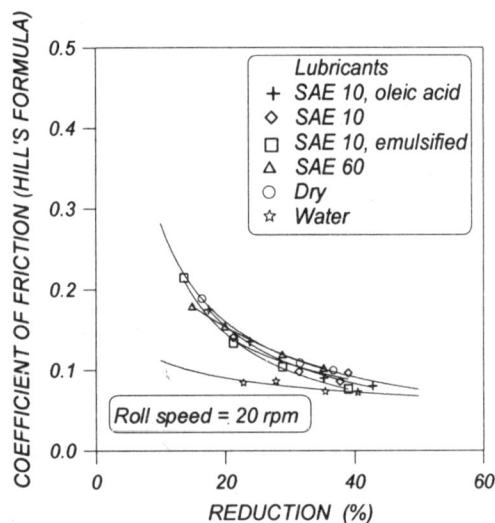

Figure 9. *Coefficient of friction at 20 rpm*

As expected, the highest coefficient of friction is observed at low speeds and dry conditions. At 20 rpm, the lowest magnitudes of the coefficient are produced by water only as the lubricant. No major differences in frictional resistance are noted when any of the oils, neat or emulsified, are used. In all cases, and as before, the coefficient of friction is reduced as the reduction is increased.

Rolling in the dry condition resulted in frictional values that are among the highest, but surprisingly not *the* highest.

There is a dependence of the frictional resistance on the viscosity at 160 rpm. The most viscous oil yields the lowest μ and the highest values are obtained when rolling dry. With oils, frictional resistance is highest with SAE 10, containing the oleic acid additive, but still significantly lower than in dry rolling, as expected. The SAE 10, neat or emulsified, leads to friction values that are practically identical and not very much different from SAE 10 and the boundary additive. The coefficient of friction reduces with increasing reduction. The coefficient decreases as the speed increases, under both dry and lubricated conditions.

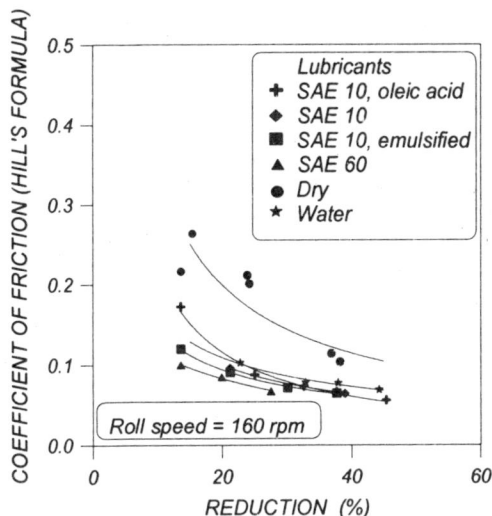

Figure 10. *Coefficient of friction at 160 rpm*

3.5. *Cold rolling of aluminium strips, using neat oils*

Aluminium strips are rolled, using mineral seal oil with lauryl alcohol, lauric acid, stearyl alcohol and stearic acid as the boundary additives, in 1, 3 and 5% (v/v) concentrations [LEN 98]. The reduction and the speed are varied and the roll forces and the torques are monitored. The coefficient of friction is obtained by the 1D model, referred to above, by matching the roll force, roll torque and the forward slip. The lowest coefficient of friction is produced by the lauryl alcohol, except at low speeds and high concentrations when use of the stearyl alcohol creates similar data. The results are shown in Figures 11 and 12.

4. Discussion

The coefficient of friction decreases with the rolling speed in all instances. It falls with increasing reduction during cold rolling of steels and increases during hot rolling of soft aluminium. When aluminium is cold rolled using mineral seal oil and various boundary additives, the coefficient falls or rises with increasing loads.

Analysis of the effects of the parameters of rolling on μ may build on the fundamental phenomena, based on adhesion and the origin of frictional resistance, resulting from the need to separate the adhesive bonds at the contact. The hypothesis is applied directly in dry rolling where the new, flattened asperities are close enough for the bonds to form. In the experiments rolling steel with steel rolls, the affinity of the metals is high and the bonds thus created are quite strong. As the reductions increase, the metal's strength increases. The true area of contact depends on the ease with which the metal deforms plastically; the harder the metal, the less the asperities will deform. The data show that μ decreases as the reduction increases at both low and high speeds and the effect of strain hardening overwhelms the rate at which new bonds are created.

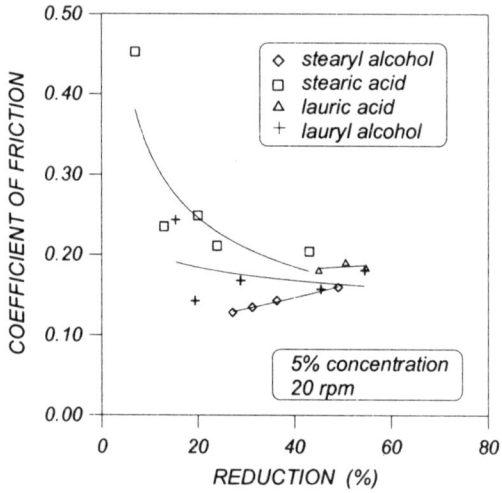

Figure 11. *Coefficient of friction as a function of the reduction and the additives; 20 rpm*

The fall of μ with increasing loads when rolling harder metals is aided by the fact that as the pressures increase beyond the yield strength, the coefficient must drop, since the shear stresses cannot rise above the shear yield strength of the metal. This limit, however, has not been reached when rolling with lubricants was performed. When rolling strips of much softer aluminium [KAR 85] friction increased with the reductions.

The viscosity of the lubricant and its pressure and temperature sensitivity affect the interactions, as does the volume of it entering the contact zone. As the speed is increased, more lubricant is dragged in to coat the surfaces and the coefficient of friction decreases. As the pressure increases so does the viscosity, and as the viscosity grows, frictional resistance falls [SCH 83]. As the reduction increases the interfacial temperatures rise, leading to lower viscosities and higher frictional resistance. The pressure effect appears to be more dominant than these when steel is rolled but the temperature effect governs when soft aluminium is rolled hot or cold. During cold rolling of soft aluminium, use of different boundary additives appeared to affect the temperatures at the contact, see Figures 11 and 12 [LEN 98]. Some of the additives enhanced the increase of the surface temperatures while the others allowed some cooling.

In one of the sets of tests on steels – see Figures 9 and 10 – another boundary additive, oleic acid, was used in the SAE 10 lubricant, in 5% concentration (v/v). The role of the boundary additive is to help in the creation and the maintenance of a thin film of the oil during contact. When rolling steels, the interfacial pressures are significantly higher than for the aluminium strips and the beneficial effect of the added oleic acid has not been observed. It is probable that the use of an extreme pressure additive would have aided the role of the oleic acid.

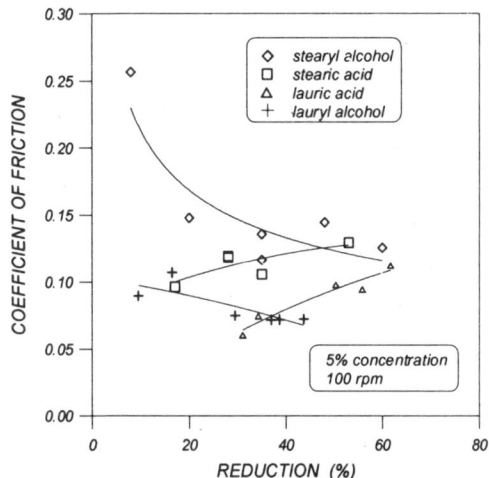

Figure 12. *Coefficient of friction as a function of the reduction and the additives; 100 rpm*

Use of the emulsion brings in further parameters. The droplet size, its standard variation within the spray, its concentration, the emulsifier and its concentration are all involved in the process and in the determination of the nature of the lubricating mechanisms. These mechanisms are the droplet-capture, which depends on the adhesion between the droplet and the moving surfaces, and plate-out. There is probably no time for plate-out to occur [NAK 88]. There is some indication that the oil droplets entered the contact zone since use of the emulsions created conditions very similar to those of the other lubricants.

Use of the water only caused very low coefficients of friction – the lowest at low speeds and among the lowest at the higher speed. The viscosity of water is very much lower than that of the light mineral seal oil. As well, water is incompressible. The coefficient of friction data indicate the probability of nearly complete separation of the rolls and the rolled metal, almost as in a hydrodynamic condition.

The interactions of the phenomena determining the nature of the mechanisms in the contact zone are illustrated in Figures 13 and 14. The effects of the relative velocity on the process are shown in Figure 13 while that of the reduction are given in Figure 14. During a particular rolling pass the following competing mechanisms are active:

– the rate at which the pressure on the lubricant increases;

– the rate at which the viscosity of the oil increases, leading to lower friction;

– the rate at which the number of contacting asperities grows, leading to higher friction;

– the pressure at which the lubricant layer breaks up, leading to higher friction;

– the relative velocity and the amount of lubricant drawn into the contact region;

– the orientation of the grooves formed by the asperities, aiding or impeding the spread of the lubricant within the contact zone;

– the growth of the bite angle, leading to more oil in the roll gap and

– the increasing surface temperature with increasing loads, leading to lower viscosity and thus, higher friction.

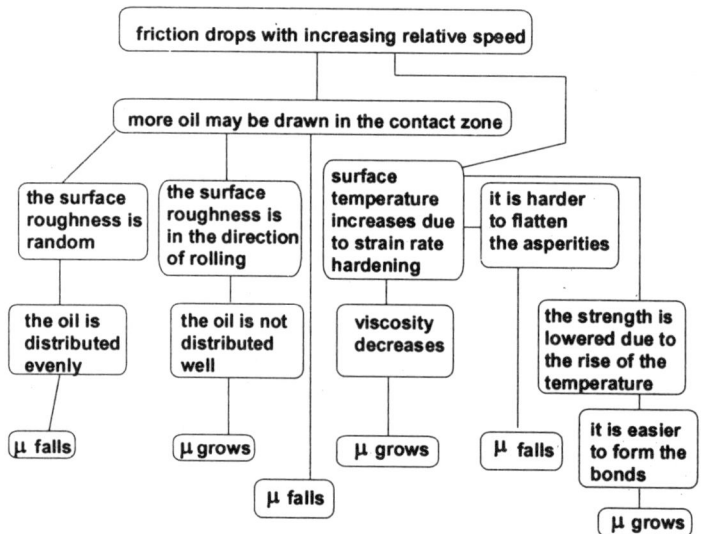

Figure 13. *Effect of the interfacial relative velocity on the tribological mechanisms in the flat rolling process*

These phenomena are connected to those shown in Figure 14 below, detailing the interconnections of the mechanisms affecting friction as the reduction is changing. If numerical data on these effects were available, a predictive relationship for the coefficient of friction may be developed. At this point in time that relationship is not yet obtainable.

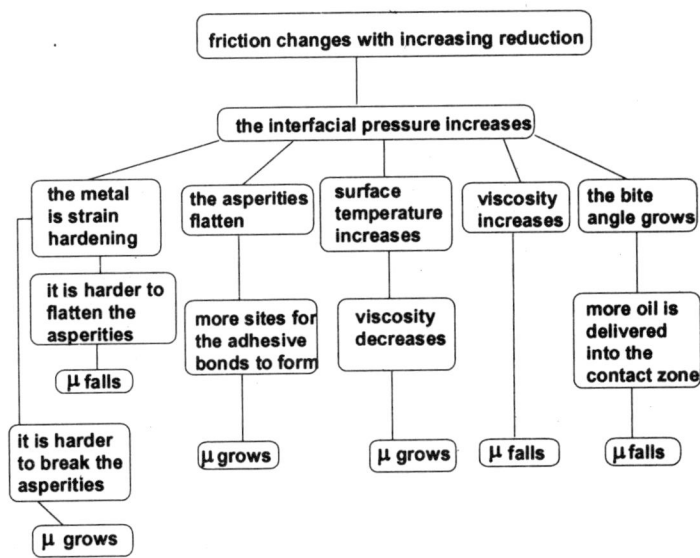

Figure 14. *Effect of the reduction on the tribological mechanisms in the flat rolling process*

5. Conclusions

The use of the embedded transducer–pin combinations was shown to yield accurate values of the coefficient of friction. These values depend on the process and material parameters and it is the interactions of the parameters that determine their magnitudes. The coefficient of friction decreases with increasing speeds. It increases or decreases with increasing reductions, depending on the interaction of the material's strength, the lubricant's viscosity and its thermal properties and the roughness of the contacting materials.

The coefficient of friction during dry rolling of steel strips is in the range of 0.1 to 0.3. When neat oils are used at higher speeds, it drops to 0.06 to 0.15. Using emulsions appears not to affect the frictional resistance. When cold rolling soft aluminium with various lubricants, μ varies from a low of 0.05 to as high as 0.4. In general, under similar circumstances, the softer the metal, the higher the coefficient of friction. Hot rolling soft aluminium with emulsions indicated frictional coefficients in between 0.05 and 0.25.

Acknowledgements

The author is grateful for the financial assistance of the Natural Sciences and Engineering Research Council of Canada, Imperial Oil and NATO.

6. References

[HOF 53] HOFFMAN O. and SACH G., 1953, *Introduction to the Theory of Plasticity for Engineers*, McGraw-Hill. Inc., New York.

[HUM 96] HUM B., COLQUHOUN H.W. and LENARD J.G., "Measurements of Friction During Hot Rolling of Aluminum Strips", *J. Mat. Proc. Techn.*, vol. 60, p. 331-338, 1996.

[JES 91] JESWIETJ., "A Friction Sensor for Sheet Metal Rolling", *Annals of the CIRP*, vol. 40, p. 231-234, 1991.

[JES 00] JESWIET J., WILD P. and SEFTON H., "Sensing Friction: methods and devices", *Int. Workshop on Friction and Flow Stress in Cutting and Forming*, Paris, p. 111-118, 2000.

[KAR 85] KARAGIOZIS A.N. and LENARD J.G., "The Effect of Material Properties on the Coefficient of Friction in Cold Rolling", *Proc. Eurotrib 85*, Lyon, p. 2-7, 1985.

[LEN 93] LENARD J.G. and MALINOWSKI Z., "Measurements of Friction During Warm Rolling of Aluminum,", *J. Mat. Proc. Techn.*, vol. 39, p. 357-371, 1993.

[LEN 97] LENARD J.G. and ZHANG S., "A Study of Friction During Lubricated Cold Rolling of an Aluminum Alloy", *J. Mat. Proc. Techn.*, vol. 72, p. 293-301, 1997.

[LEN 98] LENARD J.G., "The effect of lubricant additives on the coefficient of friction in cold rolling", *J. Mat. Proc. Techn.*, vol. 80-81, p. 232-238, 1998.

[LIM 84] LIM L. and LENARD J.G., "Study of Friction in Cold Strip Rolling", ASME, *J.Eng.Mat.Techn.*, vol. 106, p. 139-144, 1984.

[LIN 91]LIN J.F., HUANG T.K. and HSU C.T., "Evaluation of Lubricants in Cold Strip Rolling", *Wear*, vol. 147, p. 79-91, 1991.

[MCC 00] MCCONNELL C. and LENARD J.G., "Friction in Cold Rolling of a Low Carbon Steel with Lubricants", in print, *J. Mat. Proc. Techn.*, 2000.

[RAB 65] RABINOWICZ E., *Friction And Wear of Materials*, John Wiley & Sons Inc., 1965.

[ROB 97] ROBERTS C.D., "Mechanical Principles of Rolling", *Iron and Steelmaker*, vol. 24, p. 113-114, 1997.

[ROO 57]VAN ROOYEN G.T. and BACKOFEN W.A., "Study of Interface Friction in Plastic Compression", *J. Iron and Steel Inst.,* vol. 186, p. 235-244, 1957.

[SCH 83] SCHEY J., *Tribology In Metalworking,* ASM, 1983.

[SHI 00] SHIRIZLY A. and LENARD J.G., "Emulsions versus Neat Oils in the Cold Rolling of Carbon Steel Strips", in print, ASME, *J. Trib.,* 2000.

[SIE 33] SIEBEL E. and LUEG W., "Investigation into the Distribution of Pressure at the Surface of the Material in Contact with the Rolls", *Mitt. K. W. Inst. Eisenf.,* vol. 15, p. 1-14, 1933.

[STE 83] STEPHENSON D.A., "Friction in Cold Strip Rolling", *Wear,* vol. 92, p. 293-311, 1983.

[UND 50] UNDERWOOD L.R., *The Rolling of Metals,* John Wiley and Sons Inc., 1950.

Chapter 3

Friction in Modelling of Metal Forming Processes

F. Klocke and H.-W. Raedt
WZL – RWTH Aachen, Germany

Introduction

In modelling metal forming processes, whether analytically or with the help of numerical approaches, e.g. the finite element method (FEM), the input data determines the results. Therefore, a lot of effort is put into the acquisition of quantitative modelling data [PHI 93]. The measurement of the stress-strain-curves of metals is one of the important tasks. In conventional metal forming processes, up to strain rates of 300 sec^{-1} a lot of data can be accessed through various sources. The acquisition of stress-strain properties relevant to the metal cutting processes is still in its research state. Thermal bulk properties of materials can be measured exactly as well and are accessible.

The properties of the contact, however, are far less understood and less data is available. This applies to thermal problems where the heat transfer between contacting bodies is strongly dependent on contact pressure as well as on the properties of the surfaces.

This applies even more, however, to the mechanical side of the contact. The solution of the variational problem of metal flow, that can be solved using the finite element method, as well as analytical approaches, take the frictional forces into consideration. The FEM can incorporate quite sophisticated formulations of the dependency of the friction stresses on surface or bulk material properties [KOB89]. In most cases, however, Coulomb's friction law or the law of constant shear serve as a basis for the implementation. This is mainly due to the fact that the basis of measurement of frictional properties, if available at all, is these laws.

Frictional properties are most dependent on the surface roughness and further surface properties [BAY 97]. Additionally, the sliding velocity and sliding distance ('friction history') influence the friction forces. First attempts to incorporate these considerations into measurement and simulations have been made [SCH 99], but these are far from being implemented into mainstream modelling activities.

This paper presents different examples in the field of cold forming. Simulations have been used in order to get more information about the tribological contact in cold forming operations. Additionally, examples are presented in order to demonstrate the influence of the frictional properties on the simulation results.

1. Calculation of contact pressures

The effectiveness and economics of a metal forming operation is, besides other factors, determined by the wear of the tools, as well as by the environmental impact of the process.

Currently, the lubrication in cold forming is quite aggressive. In cold forging, a workpiece pretreatment is used. The workpiece is coated with a zinc phosphate layer that bonds closely with the ferrous bulk material. The zinc phosphate layer then reacts with a sodium soap, forming zinc soap, which adheres to the workpiece surface very well. The zinc soap itself can serve as a lubricant during the cold forging process. In most cases, an additional lubricant, either a solid lubricant like MoS_2, or mineral oil, is applied as well [BAY 94].

In fine-blanking, workpiece pre-treatment is not possible. Lubrication is applied through mineral oils with a high additive content. The fluid is applied through brushing or spraying onto the sheet surface. For difficult fine-blanking processes, chlorinated oils are still in use [REI 98].

The use of hard coatings on the tools of cold forming processes is intended to overcome these problems: On the one hand, hard coatings can reduce the wear of the tools and thus increase tool life. This has been shown to be effective in numerous applications in metal forming. On the other hand, hard coatings can allow for the reduction of environmentally hazardous additives in lubrication or may render the workpiece pretreatment in cold forging unnecessary. Therefore, hard coatings can lead to more environmentally friendly production. In order to further develop hard coatings or coating systems with special regard to cold forming processes, it is necessary to know the tribological contact in great detail. The mechanical and thermal loads on the surface have to be evaluated. Besides analytical approaches, the finite-element method can be used. In a coupled simulation, it allows for taking most of the physical phenomena in the tribological contact into consideration. Additionally, the dependence of physical values on input or solution values (like temperatures) can be taken into account.

In order to evaluate the tribological system in cold metal forming processes, the local sliding velocity and the local contact pressure between the plastic workpiece and the elastic die need to be known. The finite element simulation of the forming process is a powerful tool in obtaining these data. Nevertheless, the basic rules of simulation apply: only the correct input data will lead to correct results. On top of this, different aspects of finite element modelling have a big influence on the results: the mesh density and the material law (plastic or elastic-plastic) used for the workpiece will lead to different results.

Numerous process properties have an influence on the results of the forming operation. This includes the macroscopic geometry of the workpiece and dies, the material properties as well as frictional constraints. However, some of the properties cannot be incorporated in the simulation on the macroscopic scale. The surface roughness of the contacting bodies, or the coating of the tools can only be represented in models that only cover smaller dimensions. Then again, simulations on smaller scales need the results of the macroscopic simulations, e.g. the macroscopic contact stresses, depending on the geometry of the process, as input. Therefore a simulation sequence on three different scales has been implemented as shown in Figure 1.

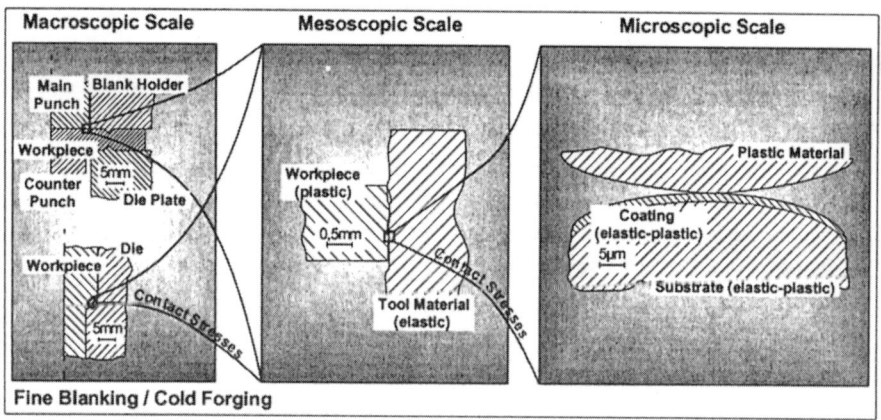

Figure 1. *Sequence of modelling in different scales*

Figure 1 shows the sequence of macroscopic process simulation, microscopic roughness simulation and the simulation of a single surface roughness asperity. These are linked in order to eventually evaluate the stresses near the surface of the tool. The contact stress results of simulations in the bigger scale are input for the simulation in the next smaller scale, respectively. As well, the temperature that is calculated in the macroscopic process simulation defines one important input to the evaluation of the tribological system. On the mesoscopic scale, the roughness of the surfaces can be represented in the simulation model. This leads to the calculation of stresses in the combined normal pressure and sliding. Finally, on the microscopic scale, it is possible to model the coating on the tool as an object in the simulation. The coating can have different properties from the substrate material. Stresses within the coating can be evaluated. With this combination of simulations on different scales, it is possible to define coating properties that are expected to lead to a prolonged tool life.

1.1. *Calculation of contact pressure in fine blanking*

The fine blanking technology has evolved from normal blanking due to the need to produce a crack-free, sheared surface with higher accuracy. Additionally, fine blanking yields a lower roll-over, less burr and better tolerances. All this is achieved by using additional tool elements. The blank holder and the counter punch induce a higher hydrostatic pressure in the shear zone which increases the forming limit of the sheared metal. The high accuracy of the part is a side effect of the more controlled metal flow.

However, the additional forces through the counter punch and the blank holder lead to additional contact stresses between workpiece and tool. Therefore, tribological problems are more difficult than in normal blanking.

An example fine blanking process has been simulated (Figure 2). Main punch diameter is 21 mm, sheet thickness is 7.5 mm. For a workpiece material with an initial yield stress of 400 N/mm^2, increasing to 1100 N/mm^2 after forming, a maximum contact pressure σ_N of 3500 N/mm^2 is calculated at the cylindrical area near the cutting edge of the main punch.

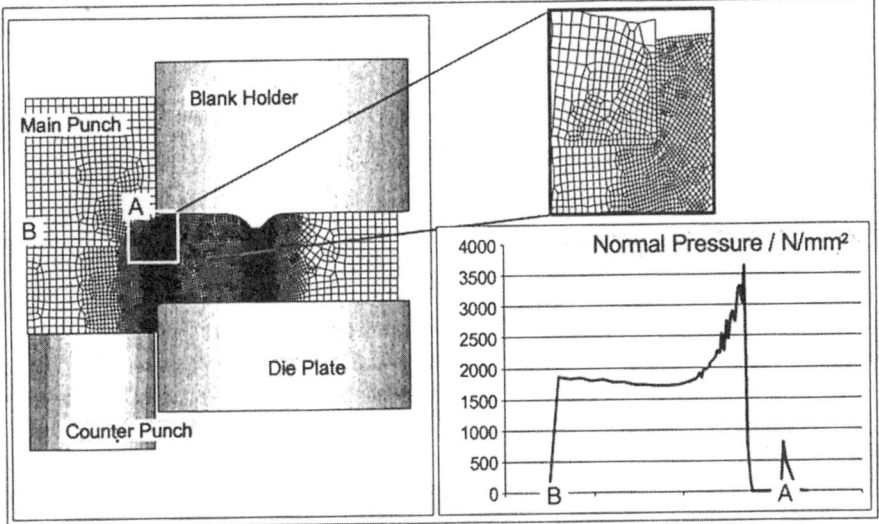

Figure 2. *Calculation of normal contact pressure in a fine-blanking operation*

1.2. Calculation of contact pressure in cold forward extrusion

Cold forging and extrusion processes have been made possible by what is nowadays a standard industrial method of workpiece pretreatment with zinc phosphate. The comparatively high yield stress of metals at cold forging temperature leads to high contact stresses between tool and workpiece.

For example, in the forward extrusion process, the maximum contact stress has been calculated to be about 2800 N/mm^2 [KLO 99] (Figure 3). The punch has a relief feature which is not visible in this figure. It is obvious that the proper description of the friction between tool and workpiece is important with respect to the results gained from these simulations. In general, higher friction forces lead to an increase of the calculated process forces. This has a direct effect on the stresses calculated between tool and workpiece, thus influencing the calculations that will be presented in sections 1.3 and 1.4.

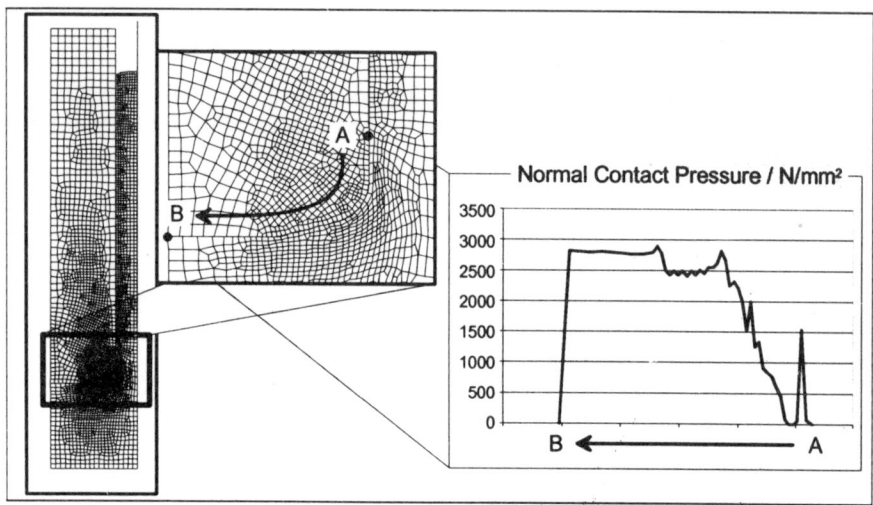

Figure 3. *Calculation of normal contact pressure in backward extrusion*

1.3. Roughness simulation

The FEM can be used to calculate elastic and plastic deformation in the mesoscopic dimension of contacting bodies. Figure 4 shows the simulation set up for such a simulation. A part of a measured roughness profile is used to build two meshes which are brought into contact with each other.

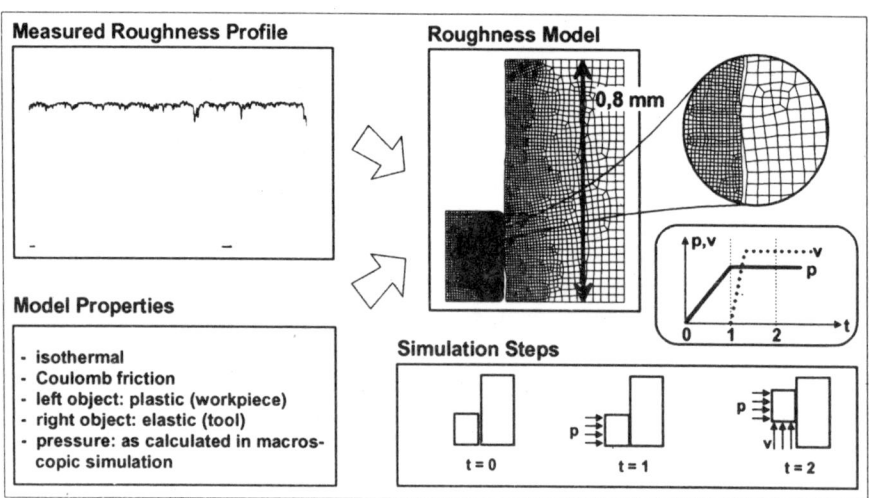

Figure 4. *Modelling roughness with the FEM*

In the initial steps of the simulation, a pressure (as calculated in macroscopic simulations) presses the left object (plastic, representing the workpiece) onto the elastic object (representing the tool). The pressure is built up in several time-steps in order to make it possible to examine the development of the surfaces in contact and the contacting stresses.

Figure 5 shows the results of the roughness modelling. The pressure acting on the plastic workpiece is much bigger than the flow stress of the material. It was taken from the slanted front of the backward extrusion punch (Figure 3) and amounts to 2500 N/mm^2. The flow of the left object's boundaries is constrained to prevent upsetting. The two objects do not touch evenly in all the possible contact areas. In cases of deeper cavities, the plastic material is not able to generate the same contact stresses in the centre of the cavities as on the peaks. Therefore, higher stresses arise at the peaks of the single asperities. Since the flow stress of the tool material is not reached at any point, it is a realistic approach to assume only elastic behaviour in the tool.

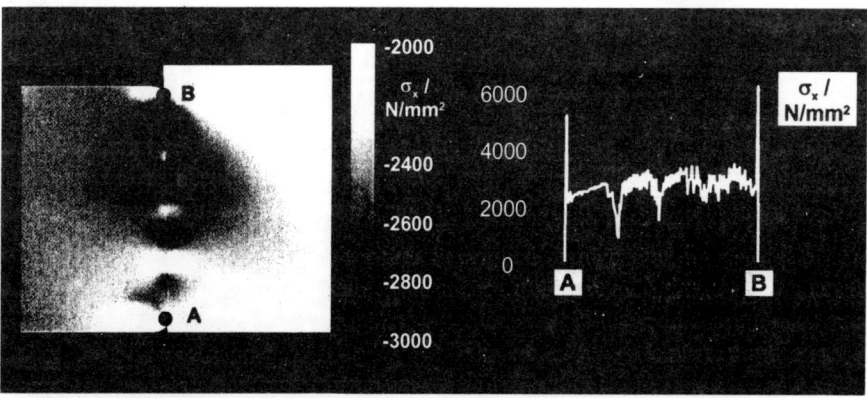

Figure 5. *Results of the roughness modelling with the FEM*

As can be seen in Figure 2 and Figure 3, the macroscopic contact stress is higher than the flow stress of the workpiece material. This is due to the high hydrostatic pressure that is generated in the metal forming operation, either backward extrusion or fine-blanking. However, the micro asperities are subject to an even higher contact stress.

1.4. Single roughness peak simulation

The results of the previous simulation are used as input for the simulation of a single surface asperity. In this simulation, only one tool surface asperity is brought in contact with an asperity of the workpiece. The workpiece material is moved into direction of the tool until the calculated normal pressure reaches the one calculated in the simulation shown in section 1.3. A hard coating on the tool material can be modelled in this scale as well (Figure 6).

With this model, it is possible to simulate the influence of coating properties on the stress state in the tribological contact. The Young's Modulus of the coating has been set to 160 GPa, 210 GPa and 260 GPa, which is respectively lower, equal to or greater than the Young's Modulus of the substrate. Consequently, the stress state in the coating varies. This is shown qualitatively in Figure 6 because the residual stresses in the coating are not included in the calculation yet. Increasing Young's Modulus of the coating in comparison to the one of the substrate, leads to an enlargement of the tensile stress field in the coating, as well as a rise in the maximum value of σ_1. This is supported by results of other researchers [BUS 98], [ROZ 89].

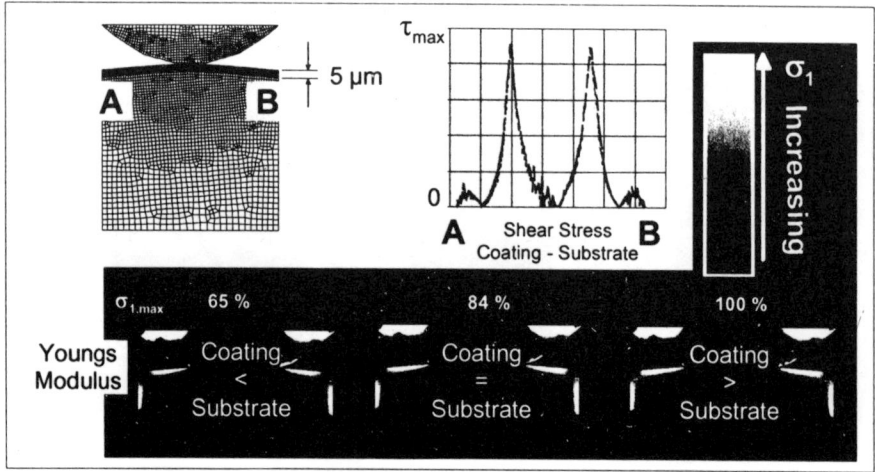

Figure 6. *Modelling of a single surface asperity with a hard coating*

At the beginning of this simulation, the coating is modelled free of any residual stresses. Current industrial hard coatings exhibit compressive stresses after the coating process. The stresses that are calculated in the simulation shown in Figure 6 are thought to be superimposed on the residual stresses after the coating process. This holds true, as coating and substrate are only subjected to elastic strain, and therefore to linear material behaviour. Although it is difficult to state absolute values of tensile stresses in the coating during the metal forming process, the trend of decreasing maximum principal stress in the coating with decreasing Young's modulus of the coating indicates a goal for coating development.

2. Workpiece geometry in fine blanking

The material flow during fine blanking is strongly controlled by the tool. Nevertheless, friction has a big influence on the resulting geometry. As a common practice in industry in cases when roll-over formation tends to result in a part that does not meet the geometric specifications, the front sides of the main punch and die plate are de-coated through a grinding operation. This leads to higher friction forces which leads to a better part geometry, even though wear resistance on the tool fronts is obviously worse. Figure 7 shows an example of how this is being examined with FEM. The friction properties between the blank and the tooling are varied in order to influence material flow. The resulting geometry clearly shows differences: the roll-over in the remaining blank does not significantly depend on the friction coefficient between the blank holder, the main punch and the blank. The roll-over in the blanked part (disk) changes much more: the roll-over height increases from 0.29 mm to 0.47 mm, while its length increases from 1.73 mm to 2.42 mm.

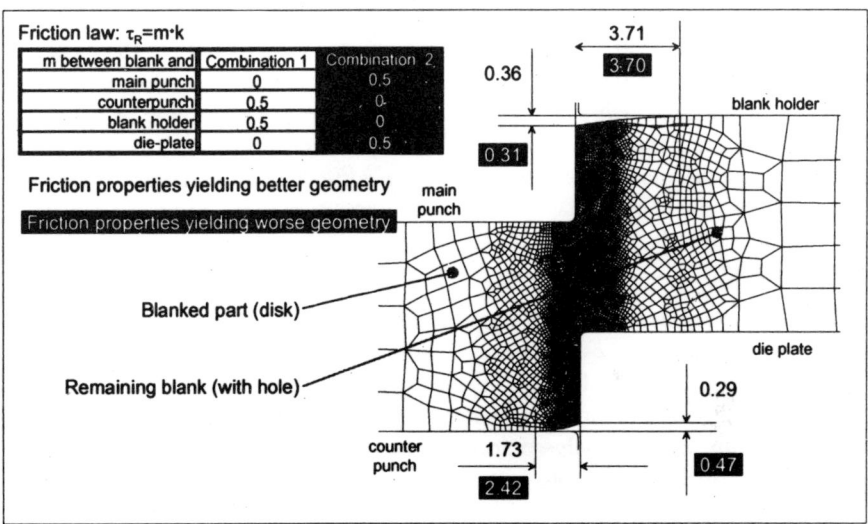

Figure 7. *Examination of workpiece geometry with varying friction properties*

Industrial experience shows that the stress-strain behaviour of the workpiece material has a big influence on the formation of roll-over as well. Materials with higher yield stress exhibit lower roll over [HAA 84]. The well-directed application of coatings with specific tribological properties in the appropriate locations opens up another possibility to improve workpiece geometry in the fine-blanking process.

3. Ductile fracture

The main aims of the application of finite element simulation of metal forming processes are the calculation of process properties like force, workpiece geometry, strain of the workpiece and stresses in tool and workpiece. This information can be used in the design stage of the forming operation in order to create an optimal process layout. Another important piece of information in the process design, for which the application of finite element analysis can be beneficial, is the appearance of ductile fracture. The process parameters of cold forging, as well as of fine blanking processes, can be optimised with respect to the formation of ductile fracture as well. Numerous criteria for ductile fracture have been proposed. Most of these criteria originated when metal forming processes were analytically treated with methods of elementary plasticity theory (e.g. upper and lower bound or slip line field methods). As the calculation of local forming behaviour was quite time consuming, the criteria for ductile fracture focussed on global characteristics of the forming process. Advanced criteria calculated the formation of ductile fracture by comparing a critical value with a "damage value" which integrated some components of the stress tensor over the equivalent strain. With the help of coefficients, these criteria could always be refitted to quantitatively predict the formation of ductile fracture in one metal forming process. However, it has been shown that by using the same coefficients and critical damage values, the prediction of fracture in a different process did not lead to sufficiently exact results [BEH 98]. This can be explained by the fact that these criteria integrate the stress-strain history into one critical value and, by doing so, neglect important information of the forming history.

The finite element method (FEM), in combination with an intelligent evaluation algorithm is expected to overcome this limitation. The simulation is capable of calculating the full forming history, i.e. the complete stress and strain tensor, with a deliberately fine resolution in geometry and time. This full forming history information can then be analysed in order to detect critical stress-strain (tensor) paths, that lead to ductile fracture.

Figure 8 shows an example of an axial compression / radial extrusion operation. The coulomb friction law with coefficients of 0, 0.1 and 0.3 has been used. In this case, the stress-strain path depicted with the development of selected elements of the respective tensors, varies only a little with the frictional boundary conditions. The influence of the frictional boundary conditions on the simulation results that are relevant for the prediction of ductile fracture, are quite small: the process force varies from 60 tons at $\mu = 0$ to about 90 tons at $\mu = 0.1$ and does not increase much more with a further increase of the friction factor. The plot of the stress path, however, shows, that the variation of the friction factor has only very little effect on the stresses during the forming process. However, in processes where different friction has an effect on the final shape of the product, or where the friction stresses change the stress state in the material for which ductile fracture should be predicted, the proper selection of the frictional boundary conditions may be more important.

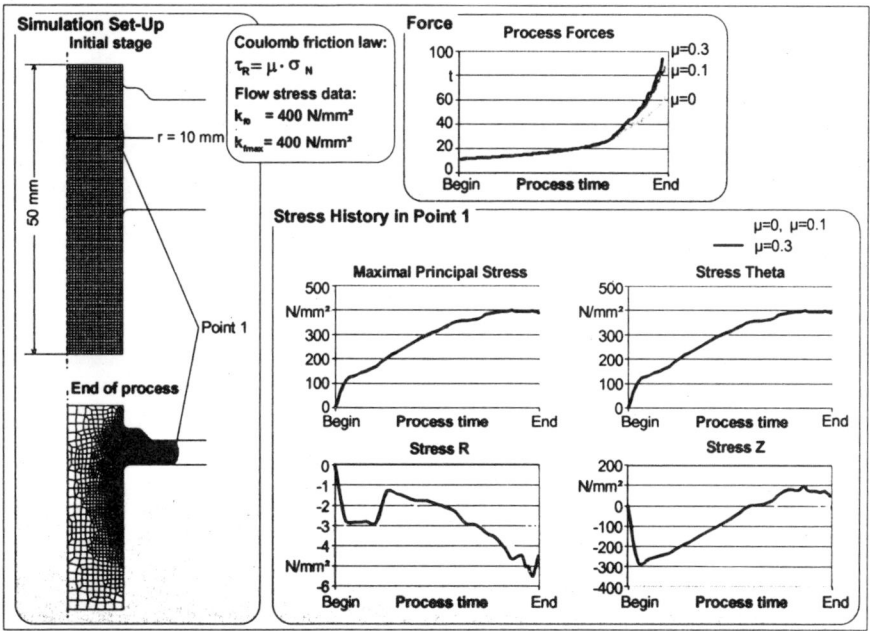

Figure 8. *Influence of frictional properties on the calculation of stress-strain paths*

4. Conclusion

Tribological phenomena can be examined with the use of the finite element method (FEM). Macroscopic contact stresses serve as input for the simulation of rough surfaces, leading to the simulation of a single surface asperity. This allows for the definition of coating properties needed in cold forging processes.

The importance of the friction factor has been shown in the simulation of a fine blanking process. Different friction factors (using the law of constant shear) lead to different workpiece shapes.

The prediction of ductile fracture, which is another piece of information that can be gained from a finite element simulation, is – in the case presented in this paper – less prone to changes in the frictional boundary constraints. Depending on the metal forming process, the stress-history of the metal changes only a little with different friction factors. Other processes may react much more to friction, though.

The importance of quantitatively correct input data for the execution of finite element simulations is well known. This paper shows that in general, this also applies to the input data concerning contacting bodies as well.

Acknowledgements

The work described in this paper was funded by the "Deutsche Forschungsgemeinschaft" (German Science Foundation) within the Collaborate Research Center 442 "Environmentally Improved Tribosystems".

5. References

[PHI 93] PHILIPP F.-D., Physikalische Prozeßdaten für die numerische Simulation von Warmumformverfahren, Thesis of the RWTH Aachen, 1993.

[KOB 89] KOBAYASHI S, OH, S.-I. ALTAN, T., *Metal Forming and the Finite-Element Method*, Oxford University Press, 1989.

[BAY 97] PETERSEN S. B., MARTINS P.A.F., BAY N., "Friction in bulk metal forming: a general friction model vs. the las of constant friction", *Journals of Materials Processing Technology*, 66, 1997.

[SCH 99] SCHAFSTALL H., "Neue Ansaetze zur Beruecksichtigung des Reibverhaltens bei der Simulation in der Kaltumformung", *VDI Jahrestreffen der Kaltmassivumformer*, 10-11 February 1999, Duesseldorf.

[BAY 94] BAY N., "The state of the art in cold forging lubrication", *Journal of Materials Processing Technology*, 46, 1994, p. 19-40.

[REI 98] REICHENBERG B., "Umformen und stanzen wie geschmiert", *Industrieanzeiger*, 50, 1998.

[KLO 99] KLOCKE F., RAEDT H.-W., "Examination of Tribological Tool Load in Cold Forging and Fine Blanking with the Finite Element Analysis", *6th ICTP*, 19-24 Sept. 1999, Nuremberg.

[BUS 98] BHUSHAN B., "Contact mechanics of rough surfaces in tribology: multiple asperity contact", *Tribology Letters*, 4, 1998, p. 1-35.

[ROZ 89] ROZENBERG O. A., CEKHANOV Y. A. SHEJKIN, S. E., "Wissenschaftliche Grundlagen fuer die Entwicklung verschleißfester, beschichteter Kaltumformwerkzeuge", *Wissenschaftliche Zeitschrift der Technischen Universität "Otto von Guericke" Magdeburg*, 33, 1989.

[HAA 84] HAACK J., BIRZER F., Fine-Blanking, *Practical Handbook*, Feintool AG Lyss, 1984.

[BEH 98] BEHRENS A., LANDGREBE D., "Beurteilung von alternativen Stadienplänen in der Massivumformung", *15. FEM Kongress*, 16-17 Nov. 1998, Baden-Baden.

Chapter 4

Friction and Wear in Hot Forging

Claudio Giardini and Elisabetta Ceretti
Dept of Mechanical Engineering, University of Brescia, Italy

Giancarlo Maccarini and Antonio Bugini
Dept of Engineering, University of Bergamo, Italy

1. Introduction

Every time a new piece is forged, the residual life of the dies involved in the forming operation decreases due to wear phenomena until the maximum allowed wear is reached: at this point the dies must be substituted or, if possible, reworked. To forecast the wear level is a fundamental task in cost definition for the piece produced.

Wear can be correlated to several working parameters and mainly to sliding length, normal pressure and tool hardness according to Archard's model [ARC 53].

The aim of the present paper is to analyse the influence of friction on these variables and, consequently, on die wear in a hot forging operation, particularly the stamping of a flange. The information derived from the model, once validated, can be utilised to correctly design the dies and define the process variables.

In particular the steps followed during the present research have been:

– to define a suitable wear model;

– to implement this model in an F.E.M. program;

– to validate this model with experimental tests;

– to extend the results of the model in forecasting the material behaviour varying the friction factor.

The wear model, found in literature, has already been applied in turning and extrusion, furnishing appreciable results [CER 95, PAI 95]. Particularly, wear of the dies is mainly considered abrasive due to very hard particles placed between piece and die which progressively damage the dies.

After an initial study where the results of the proposed model are compared with experimental data, in successive simulations friction has been varied in order to highlight its influence on both material flow and die wear.

The most interesting parameters investigated have been: material flow (velocity) in the billet, stresses on the dies (pressure), temperature distribution both in billet and dies (this affects the die hardness) in order to forecast the wear on the dies. The simulation results have shown how both friction and temperature affect the die wear even if it seems that each zone of the dies must be treated separately from the others.

As already said, we tried to apply the method (for validating it) and to analyse the friction influence studying an actual process, i.e. the stamping of a flange. The deformation process is divided into two successive steps: in the first step the billet is simply upset, while the second step is the forging into closed dies with flash formation. Both of them are performed with a hammer. This means that we do not have a constant descending velocity for the upper die, but a total energy value.

Due to the particular geometry of the flange, a two-dimensional asymmetrical approach has been used. To correctly represent the actual phenomenon, a non-isothermal simulation has been utilised.

The experimental tests have been conducted in co-operation with a company. Other information has been found in the literature or from actual practice or from metals handbooks [AV 1, AV 2, ALT 83].

2. Wear study

To predict die wear is a fundamental task when designing a new tool or die. In fact, the choice of the appropriate die geometry (rake angles, fillet radii, etc), die material and lubrication conditions at the die workpiece interface affect the die wear which results in bad final quality of the produced part and in die life reduction. To know the critical points for wear development will allow better die design, changing the die geometry or thermal treating the die surface or using harder material for all or part of the dies themselves, and the identification of the optimal lubrication conditions.

In the present paper the data coming from FEM simulations are used to forecast the die wear and to understand the role played by friction in wear development.

The simulation code is a commercial one and it is a Lagrangian two dimensional finite element program (DEFORM®) [DEF 1]. The research consisted of four main points:

1. definition of the appropriate wear model and of the experimental case study (industrial problem);

2. extraction of the simulation data and estimation of the die wear through a calculation subroutine;

3. comparison between the die wear prediction and the experimental wear profile;

4. analysis of friction influence on die wear.

3. An approximate wear model

Estimation of wear using FE simulations requires that appropriate wear models be available in a discretised form. The wear model considered has been suggested in the literature [ARC 53, FEL 80, REN 83], for tool wear prediction. This model was found to give reasonable results and, therefore, it is used in this paper.

Namely, the die wear is abrasive wear, due to the presence of harder particles (included in the surface of the workpiece material), at the interface between die and piece. These particles remove material form the die and scratch the die surfaces.

When many parts are worked, the wear phenomenon becomes relevant and the final part dimensions could be out of tolerance or the die could be damaged losing its surface treatment (Figure 1).

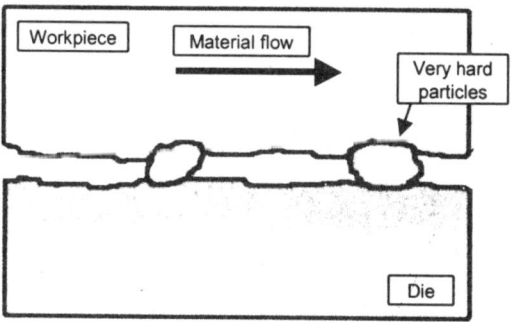

Figure 1. *Schematics of the abrasive wear phenomenon: it is due to very hard abrasive particles included in the workpiece surface at the workpiece-die interface*

The original wear model, according to [ARC 53 and REN 83], states that:

$$V = K \cdot \frac{q \cdot s}{H} \tag{1}$$

where:

V	is the wear volume
K	is a constant which depends on the material couple and the interface conditions
q	is the normal pressure acting between piece and die
s	is the length of the zone where sliding takes place
H	is the hardness of the worn die: this hardness depends on the die temperature

The model, derived from the previous formula, utilised to study the abrasive die wear, can be written as follows:

$$Z_{AB} = \int K_1 \frac{p^{a_1} \cdot v^{b_1} \cdot dt}{H_d^{c_1}} \tag{2}$$

where:

Z_{AB}	is the abrasive wear depth
p	is the local pressure
v	is the local sliding velocity

dt	represents the incremental time interval
H_d	is the tool hardness
K_1, a_1, b_1, c_1	are experimental constants.

The integral is extended from the beginning to the end of the forming process and it must be considered as cumulative wear. Obviously this wear depth must be considered locally on the die because the several figures involved differ from one point to each other.

For prediction of wear, the exponent values a_1, and b_1 are commonly taken to be equal to unity while c_1 is taken equal to 2 (for steel). These exponents can be modified by the user in the wear program, though for fine tuning purposes.

The coefficient K_1 simply scales the magnitudes of the predicted wear profiles, and it is found experimentally by comparing the predicted wear profiles with the actual measured wear profiles [FEL 80].

It is essential to note that the abrasive wear model always predicts material removal from the tools.

In previous work this approach has already been applied trying to identify the several figures of expression (2) all along the deformation process [HAN 90].

4. Implementation

Once the wear model has been chosen, a subroutine for automatic wear estimation has been implemented.

The wear model allows us to evaluate the wear of each point of the die surface through a time integral, extended along all the process duration, as a function of hardness, local pressure, relative velocity between piece and die and temperature.

The expression proposed has been rearranged for our study according to the following:

$$Z_{AB} = K_1 \cdot \sum_{j=1,n} \frac{p_j^{a_1} \cdot v_j^{b_1}}{\left(H_d^{c_1}\right)_j} \cdot \Delta t_j$$

[3]

for each mesh node of the simulated dies (Figure 4). In expression (3) we have:

Z_{AB}	is the abrasive wear depth for one node
n	is the total number of steps
Δt_j	is the time interval
p_j	is the pressure
v_j	is the node sliding velocity

H_d is evaluated once the temperature of the node is known according to the graph reported in Figure 6. All of these figures refer to the j-th step of the simulation.

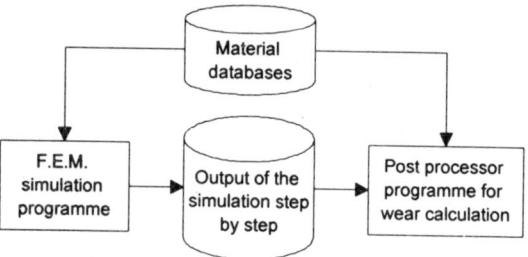

Figure 2. *Schematics of how the method is implemented in our calculation environment*

The developed programme (post processor of the FE analysis) is able to extract all these figures from the simulation database and calculates, for each step and for each node of the two dies, the incremental Z_{AB} values from the beginning to the end of the process. The schematics of the method implemented is reported in Figure 2.

5. The case study and the model validation

The case study, developed in co-operation with a company, refers to the industrial hot forging of a flange between closed dies (Figures 3 and 4). The production steps of the flange are: heating of the billet to 1200°C, forging between flat parallel dies, stamping in closed die with flash formation, removal of the flashes.

The two deformation steps are performed with a hammer: this means that the FEM programme does not work with a constant descending velocity for the upper die, but it bases its simulation starting from the total energy. In order to do this the *"hammer energy"* option has been activated. This energy has been evaluated on the basis of the hammer mass plus the die mass, the falling height, the gravity acceleration and the process efficiency (η).

$$E = \left(M_{hammer} + M_{die}\right) \cdot h \cdot g \cdot \eta$$

[4]

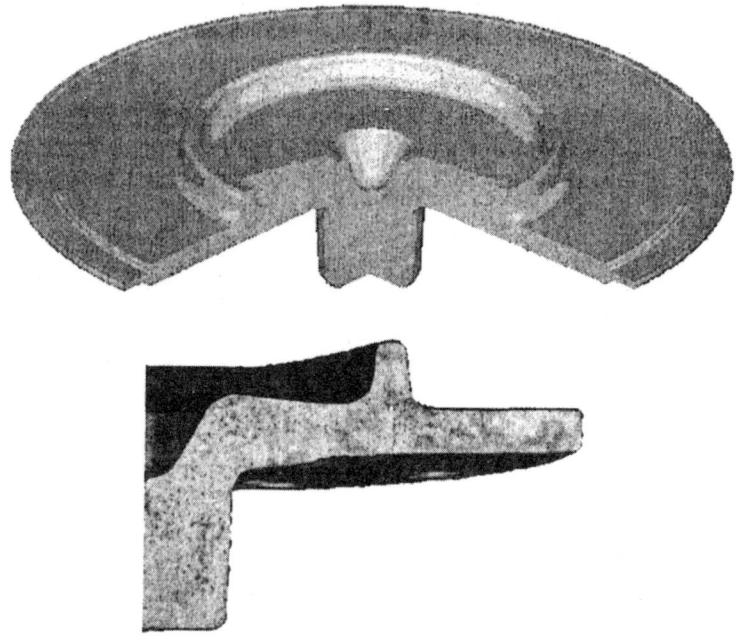

Figure 3. *The workpiece studied: the model and the actual piece*

The forging and stamping phases are realised with a hammer of 4000 Kilos (including the dies weight). All the data for the FEM model implementation come from industry: dies and billet geometries, flow stress for billet (S355J2G3) and dies (H13), working temperature (1200°C for the billet and 300°C for the dies) and hammer energy and efficiency.

The FEM model is axi-symmetrical, the simulation type is non isothermal. Friction at the die work-piece interface has been modelled as shear, with a constant value of m = 0.35.

The two forging steps have been simulated and attention has been focused on the stamping step which is critical for the die wear analysis (Figure 4). The simulation results, in terms of material flow, pressure acting on the dies, piece and dies temperatures, have been extracted and implemented in the wear calculation subroutine.

In order to compare the ability of the simulation procedure to represent the actual wear on the dies, some experimental measures have been conducted by means of a probe mounted on a measuring machine. Experimental observation has shown that the most worn zones are those reported in Figures 4 and 5.

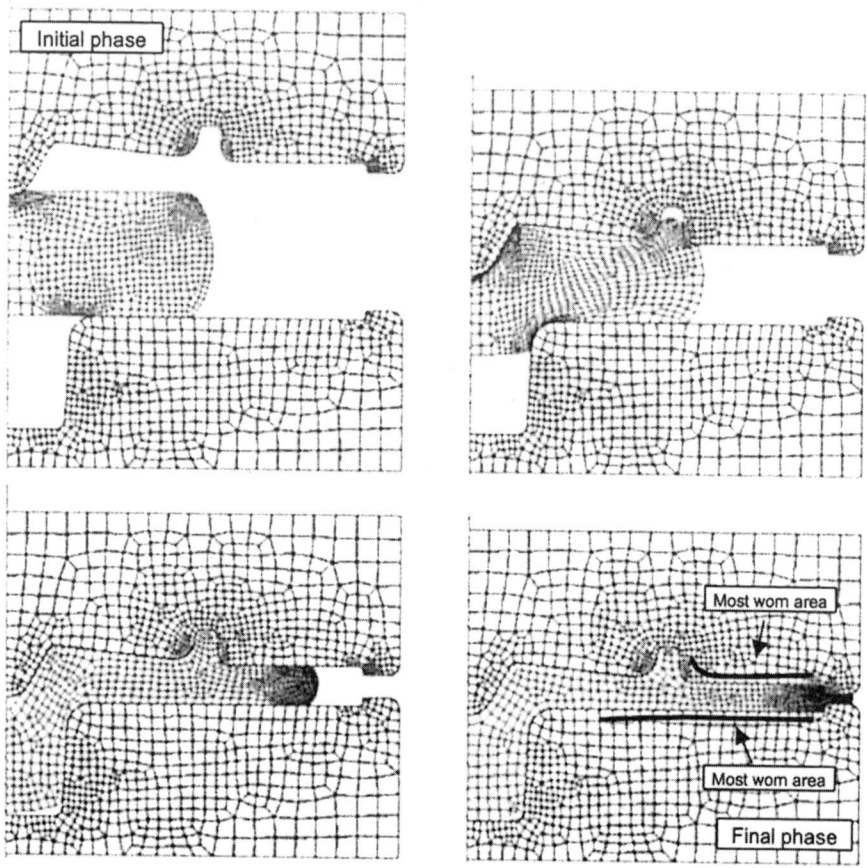

Figure 4. *Initial, intermediate and final phases of the second step of the deformation process (m = 0.35); the most worn zones from experimental evidence are highlighted*

In this way it has been possible to sample the wear level on a predetermined point of the dies profile. By executing the calculus reported in paragraph number 4, the theoretical wear has also been evaluated. As already said in paragraph number 3, a suitable value for K_1 has been chosen in order to correctly scale the magnitude of the calculated wear.

The hardness was considered to be a function of temperature according to Figure 6 (from literature) [AV 1, AV 2].

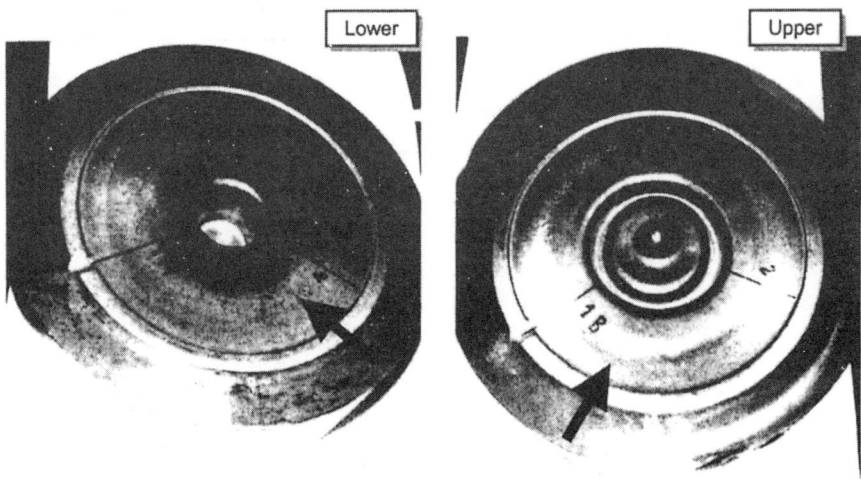

Figure 5. *Zones where the comparison between FEM results and experimental evidence are compared*

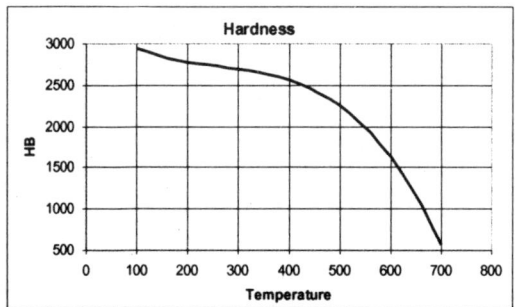

Figure 6. *Hardness vs. temperature behaviour*

Figure 7 shows the original die profiles, the measured profiles (worn dies) and the calculated wear ($m = 0.35$). It can be noticed that with a value of $K_1 = 1000$ the experimental results and the wear model results are in good agreement.

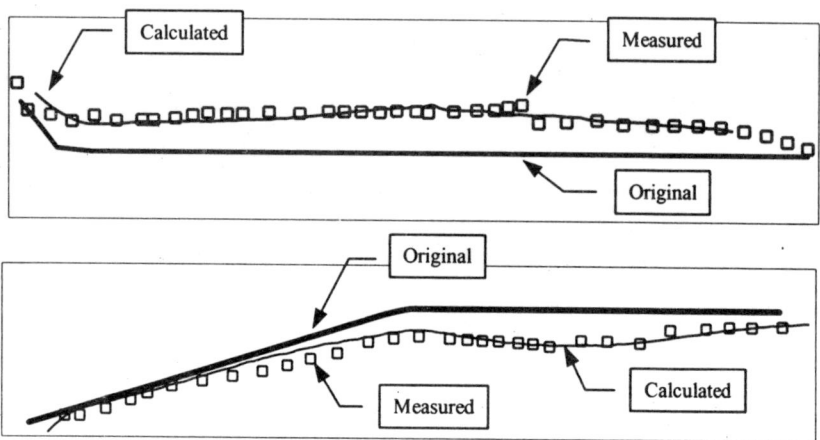

Figure 7. *Comparison between original profile, calculated and measured (experimental) wear along the upper and lower dies (m = 0.35)*

6. Friction influence on die wear

Once the model and the approach have been validated, further simulations have been conducted in order to outline the influence of the several parameters involved and to furnish important suggestions in the design of forging operations.

The two die profiles have been divided into eight areas (four for the upper and four for the lower die) according to Figure 8.

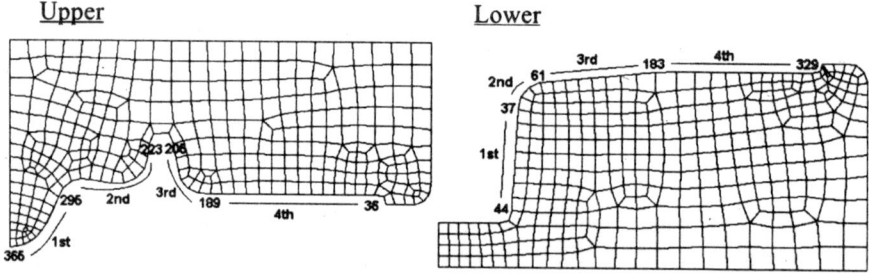

Figure 8. *Subdivision of the upper and lower dies into four different zones*

The simulations have been conducted varying the friction factor m (equal to 0.2 and 0.5). The results obtained are reported in Figure 9. This figure shows how the wear level varies locally as a function of m. The four different zones (for upper and lower dies) are represented.

By analysing the results of these simulations, it is important to note how a general behaviour could not be outlined. The most important information is:

– the most worn zones remain the same even if the friction factor varies;

– in some cases it seems that the wear level could be directly linked to the m value, but, generally speaking, there is not a direct effect.

As a consequence, we can say that:

– a single zone must be studied individually;

– an important role is played by the material flow (i.e., the process history) and the velocity field, i.e. the geometry of the dies.

This is quite evident if we refer to the wear expression chosen (formulas 2 and 3). In fact, while it is clear how the single contribution Z_{AB} reported in formula (3) directly depends on p, v and H values, it is not the same for m. In fact, we can expect higher p values and higher temperatures (and consequently lower H) as friction increases.

On the other hand v has a completely different behaviour as the compression takes place:

– in the initial phase, the material flows with lower "centrifugal" velocity while the die cavities are filling;

– in the final phase, the die cavities being already filled, the material is obliged to increase its radial velocity.

This is confirmed by Figure 10 where the velocity distribution and the piece deformation are reported.

These figures report differences in the plastic flow when considering different friction factors. This means that not only the several figures investigated (local velocity, local pressure, hardness) are different during the compression process, but also the history in the plastic flow, i.e. the integration extremes and, consequently, the time steps Δt. Note that at equal time (in the examples 0.826 sec.) the strokes are different for the three cases. This means that each zone must be studied independently one from each other.

Another confirmation is given in Figure 11 where the behaviour for the figures involved in wear definition are plotted for a given point (one of the most worn) as the deformation process takes place. It is possible to see how when the friction increases also the pressure grows. The opposite is for hardness being inversely dependent on temperature. Finally, the velocity behaviour is also confirmed: in the first phase we have higher velocity at low friction while in second phase the behaviour becomes the opposite having higher sliding velocity and high friction.

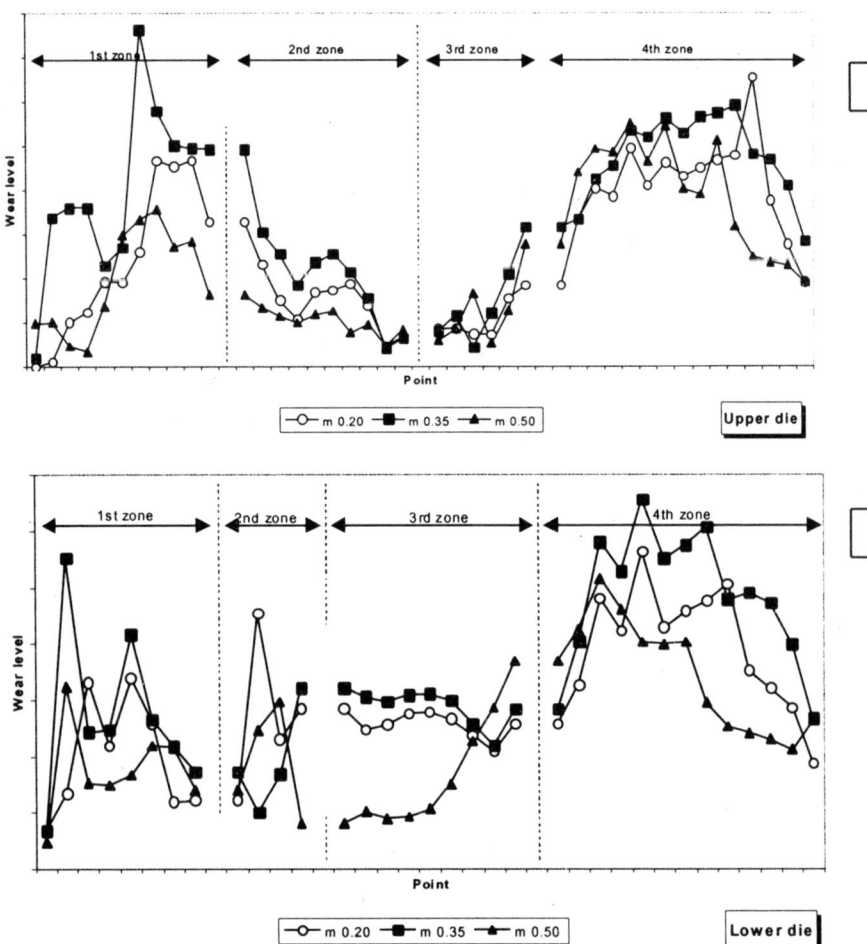

Figure 9. *Comparison between the wear level along the several zones of the upper (A)and lower (B) dies as function of friction factor*

7. Conclusions

In the present paper an analysis on the wear level as a function of the friction factor *m* has been carried out. From the validated model (compared with experimental tests) a general behaviour has been investigated. The results of this research are reported in chapter 6.

First of all the wear model has proved to be valid. In fact, the experimental worn profile matches perfectly the results obtained with the analytical model. This is important for the die design. Being able to forecast the wear profile will enable one to choose the appropriate die geometry and working conditions improving the dies life and the part quality

In conclusion we can say that for some zones the wear increases with m, while in others it decreases. This means that, first of all, we have to limit our investigation only to where the wear is higher and, secondly, that an increase in friction does not correspond to a general increase in wear level.

This is also confirmed from previous work [HAL 90] where it has been shown how less wear is found when relatively rough surfaces for the dies are considered.

An important role is also played by the plastic flow history in as much as the sliding length and velocity greatly depend on it.

Moreover, the validity of the FEM approach for process optimisation has been proved. This gives the designer the possibility of forecasting quite correctly the wear level reached during the forming process in order to project them minimising the maintenance operations on the dies.

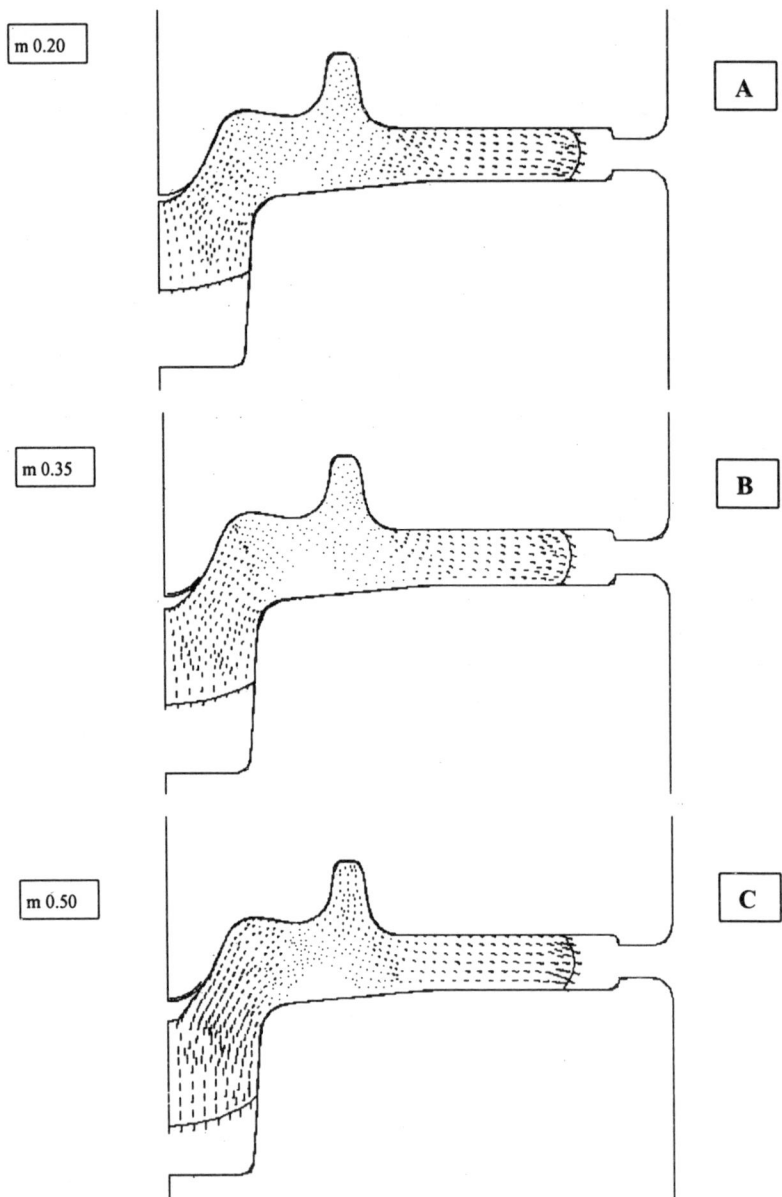

Figure 10. *Differences in strokes, material flows and velocities at the same time step for the three considered cases (**A** m = 0.20, **B** m = 0.35, **C** m = 0.50)*

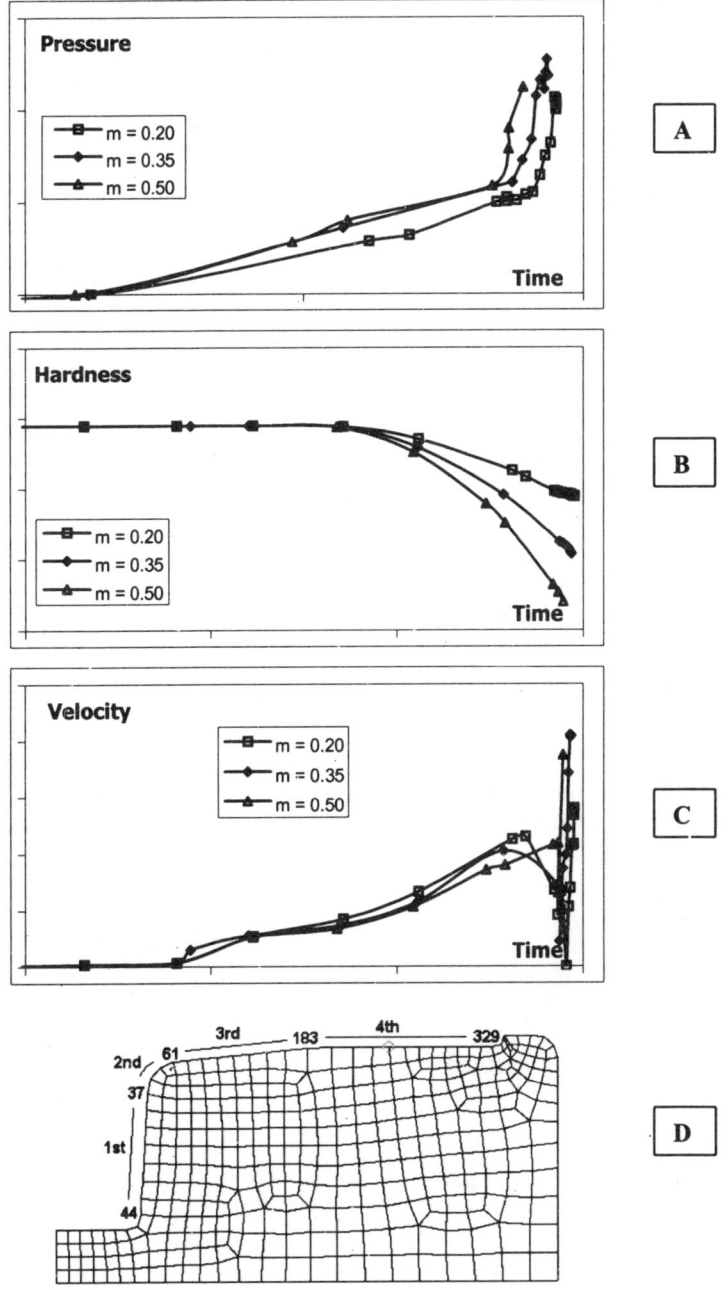

Figure 11. *Behaviours for the figures involved (A pressure, B hardness, C velocity) in the wear definition along the deformation process for a given point (one of the most worn). This point is highlighted in D*

8. References

[ARC 53] ARCHARD J.F., "Contact and rubbing of flat surfaces", *J. Of Appl. Phys.*, n° 24, p. 981, 1953.

[AV 1] AA.VV., *Metals Hanbook Vol.1. Properties and selection: Irons, Steels and High-Performances Alloys*, Tenth ed.

[AV 2] AA.VV., *Metals Hanbook Vol.18. Friction, Lubrification and Wear Technology*, Tenth ed.

[ALT 83] ALTAN T., OH SOO-IK, GEGEL H.L., *Metal forming. Fundamentals and application*, American Society for Metals ed., 1983.

[CER 95] CERETTI E., FALLBOEHMER P., WU W.T., ALTAN T., Simulation of high speed milling: application of 2D FEM to chip formation in orthogonal cutting, ERC Report NO. ERC/NSM-D-95-42, 1995.

[DEF 1] DEFORM, Scientific Forming Technologies Corporation, *Deform-2D User's Manual*, Columbus, OHIO.

[FEL 80] FELDER E., MONTAGUT J.L., "Friction and wear during hot forging of steel", *Tribology International*, p. 61-68, 1980.

[HAN 90] HANSEN P.H., BAY N., "A flexible computer based system for prediction of wear distribution in forming tools", *Adv. Techn. of Plasticity*, 1990, p. 19, Kyoto, 1990.

[PAI 95] PAINTER B., SHIVPURI R., ALTAN T., Computer-aided techniques for the prediction and measurement of die wear during hot-forging of automotive exhaust valves, ERC Report NO. ERC/NSM-B-95-06, 1995.

[REN 83] RENAUDIN J. F., BATIT G., THORE Y., FELDER E., "A way to build a computer program for forecasting the abrasive wear of hot forging dies", *11th Int. Drop Forg. Conf.*, Köln, 1983.

Chapter 5

Basic Aspects and Modelling of Friction in Cutting

E. Ceretti
Dept of Mechanical Engineering, University of Brescia, Italy

L. Filice and F. Micari
Dept of Mechanical Engineering, University of Calabria, Italy

1. Introduction

The role of machining process modelling is recognised in industry, due to the relevant advantages that an effective and reliable theoretical model can supply. Such a model would in fact represent a very powerful tool to design cutting tools and to select the cutting conditions, dramatically reducing the amount of trial and error expensive experimental tests, which are normally carried out for this purpose.

Within this framework the potentialities linked to the use of advanced numerical models and in particular finite element techniques have been recognised by a large number of researchers all over the world [STR 85], [STR 90], [SHI 95], [MAR 95], [VAZ 98]. Nevertheless, it is worth pointing out that up to now no general and reliable models have been developed and that the industrial applicability of FEM techniques for a quantitative analysis of machining presents large difficulties. In particular, the characterisation of material behaviour at the strain, strain rate and temperature which occur during the cutting process, the definition of the frictional conditions at the chip-tool interface and the development of effective chip separation criteria probably represent the most relevant and still partially unsolved problems.

In the paper attention is focused on some basic aspects of friction modelling in cutting and in particular on the influence of friction modelling on the effectiveness of the numerical simulation of the process. The paper is subdivided in two different sections. In the former some of the most important friction models up to now proposed in the technical literature are taken into account and utilised to carry out numerical simulations of a typical orthogonal cutting process characterised by continuous chip. Attention was focused on the analysis of the influence of friction modelling on some of the most important results typically supplied by a numerical simulation of machining. In particular the predictive capability of the cutting force, the chip contact length, the chip thickness and the shear plane angle were analysed comparing the numerical data with the results of some experimental tests.

In the latter part of the paper, another orthogonal cutting process characterised by segmented chip formation is discussed. A reliable chip separation criterion was implemented through a proper user-subroutine and the influence of friction modelling on the prediction of chip shape was analysed.

2. Orthogonal cutting with continuous type chip

2.1. *Basic remarks about numerical simulations*

First of all any numerical simulation must be based on a suitable material characterisation to be effective. It is well known that during an industrial machining process very heavy conditions occur so far as strain, strain rates and temperatures are concerned: the material undergoes large effective strains, ($\bar{\varepsilon} \geq 1$), very high effective strain rates, ($\dot{\bar{\varepsilon}} = 10^4 \div 10^6 \text{ s}^{-1}$), and high temperatures (T>300°C).

All the tests presented in the first part of this paper refer to the orthogonal cutting process of mild steel (AISI 1035) specimens. Thus the flow stress was expressed as a function of effective strain, effective strain rate and temperature, according to the model proposed by Shirakashi, Maekawa and Usui [SHI 83] and utilised also by Lin and Lin [LIN 92]:

$$\sigma_f = A_0(T,\dot{\varepsilon})\left(\frac{\dot{\varepsilon}}{1000}\right)^{0.0195} \bar{\varepsilon}^{0.21} \quad [\text{MPa}]$$

$$A_0(T,\dot{\varepsilon}) = 1394e^{-0.00118T} + 339e^{-0.0000184\left[T-\left(943+23.5\ln\left(\dot{\varepsilon}/1000\right)\right)\right]^2}$$

The above flow stress model maintains its validity within the following variables range: $T = (293 \div 970)\ {}^\circ K$, $\bar{\varepsilon} = (0.05 \div 2)$, $\dot{\varepsilon} = (10^{-3} \div 10^4)\,s^{-1}$.

So far as friction modelling is concerned, it is well known that stress distribution on the rake face is typically non linear [ZOR 58]; [TRE 77]. Moreover, while the normal stress monotonically increases toward the tool edge, the frictional stress first increases but then saturates in the highest pressure zone close to the tool edge. These observations induced some researchers to propose the existence of two distinct regions on the rake face, namely a *sliding* and a *sticking region*. In the former zone the normal stress is relatively small and dry sliding (Coulomb) theory is still able to provide a suitable model for the phenomenon. In the latter, on the contrary, the normal stress is so large that the real contact area is equal to the apparent one and the frictional stress saturates to an almost constant value (i.e. the shear flow stress of the chip material).

A couple of friction models able to reproduce the above described experimental observations were taken into account and utilised in the numerical simulations.

The former is the model proposed by Shirakashi and Usui [SHI 73] which relates the frictional stress τ to the normal stress σ at the chip-tool interface:

$$\tau = \tau_f\left[1 - e^{-k\frac{\sigma}{\tau_f}}\right]$$

In the above equation τ_f is the shear flow stress of the chip material and k is a constant depending on the chip-tool materials combination, which allows one to fit the experimental frictional stress vs. normal stress curve on the rake face. According to the data reported by Usui and Shirakashi [USU 82], the k-value was fixed at 1.6 in the numerical simulations. Such a value, in fact, properly fits the experimental data for the workpiece-tool combination utilised in the research here addressed (work material: AISI 1035; tool material: uncoated sintered carbide – P20 grade).

The latter is a simple model which takes into account in a very straightforward way the existence of the two regions observed on the rake face, namely sliding and sticking. Thus a constant coefficient of friction, according to the Coulomb model, is utilised in the sliding region, while a constant frictional stress, equal to the shear flow stress of the chip material, is applied in the sticking region. Such a model can be expressed by means of the following mathematical formulation:

$$\tau(x) = \mu\sigma(x) \quad \text{when} \quad \tau < \tau_f$$

$$\tau(x) = \tau_f \quad \text{when} \quad \tau \geq \tau_f$$

τ and σ being the frictional and the normal stress and τ_f the shear flow stress of the chip material. It is worth pointing out that in this model the extension of the sticking region depends on the choice of the friction coefficient value in the sliding region: this value, in fact, determines the fulfillment of the equality $\tau = \tau_f$ and thus the saturation of the frictional stress. A couple of different values of μ were considered in the numerical simulations, namely $\mu = 0.5$ and $\mu = 0$.

Figure 1 reports the frictional stress vs. normal stress distributions obtained utilizing the Shirakashi and Usui model and the sticking-sliding model and assuming $\tau_f = 500$ MPa; in the latter case both the distributions for the two investigated μ values are presented.

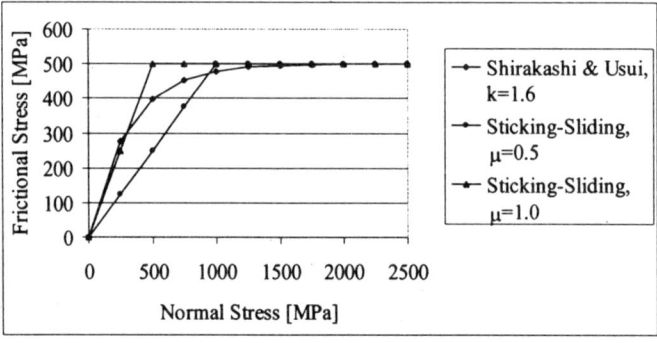

Figure 1. *Frictional vs. normal stress distribution*

In the next paragraph the numerical predictions obtained with the above described models will be compared with some experimental results. A further comparison will be carried out taking into account another well known friction model, namely the *constant shear model*. In this case the experimental observations about frictional stresses on the rake face are not taken into account so far and a constant frictional stress on the rake face is assumed, equal to a fixed percentage of the shear flow stress of the chip material: $\tau = m \, \tau_f$. According to the data reported by the authors in some recent publications [CER 97], [CER 98] and [FIL 99] two different values of m were investigated, namely m = 0.5 and m = 0.8.

2.2. Simulations and experiments

Before the simulations, an orthogonal machining experiment was performed. A tube of AISI 1035 steel with a wall thickness equal to 3.0mm was set up for machining on a lathe; the cutting tool was an uncoated sintered carbide (P20) with a rake angle equal to 6°; no lubricant was used at the tool-chip interface. The tests were carried out with two different cutting speeds equal to 95m/min (1.58m/s) and 125m/min (2.08m/s), while the feed was maintained constant and equal to 0.25mm/rev. With the above mentioned cutting conditions the continuous type chip was observed throughout the experiments. Both the cutting force and the chip thickness were measured; in the former case a piezoelectric dynamometer was utilised while in the latter the chip thickness was measured both with the well known *weight method* and using an optical microscope able to superimpose two different objects in order to estimate lengths with an approximation grade equal to 0.1mm. Figure 2 shows a typical side view of the chip.

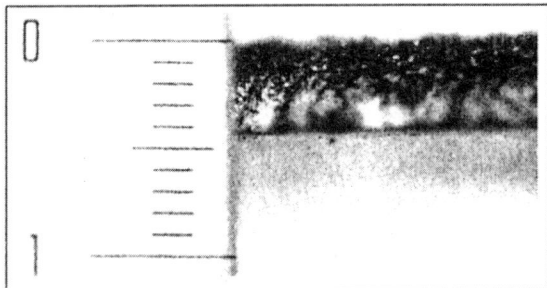

Figure 2. *Chip thickness measurement (cutting speed 2.08m/s)*

Figure 3 shows an upper view of the tool rake face after few seconds of cutting; for both the investigated cutting speeds, an accurate analysis of the surface on the above described optical microscope permitted one to distinguish the contact zone between the chip and the tool and consequently to measure the contact length.

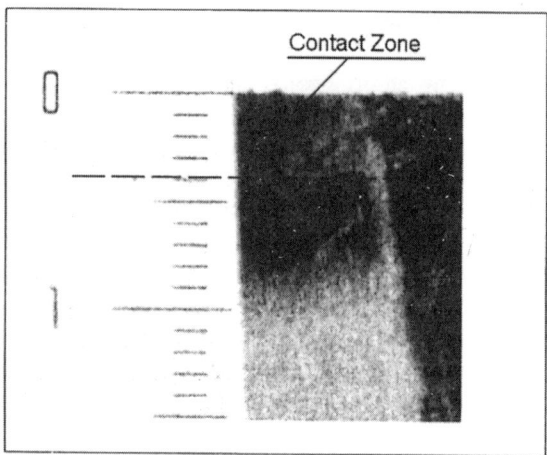

Figure 3. *Contact length measurement (cutting speed 2.08m/s)*

Table 1. *Summary of the experimental results*

Cutting Speed [m/s]	Cutting Force [N]	Chip Thickness [mm]	Contact Length [mm]
1.58	1350	0.43	0.40
2.08	1310	0.42	0.40

So far as the numerical simulations are concerned, a coupled thermal-mechanical analysis was carried out. The workpiece material behaviour at high strain, strain rate and temperature was described by means of the flow stress law above mentioned; as well the other relevant physical properties of the workpiece and the cutting tool were taken from reference [LIN 97]. Due to the cutting geometry, plane strain conditions were assumed; in the simulations the depth of cut was equal to the experimental feed (i.e. 0.25mm), while the width of cut was equal to the wall thickness of the tube.

Table 2 summarises the most relevant numerical results: in particular the predicted cutting force (F_c), chip thickness (t), shear plane angle (ϕ) and chip contact length (l_c) are reported at varying the cutting speed (V) and the adopted friction model.

Table 2. *Numerical results*

Friction model	V [m/s]	F_c [N]	t [mm]	ϕ	l_c [mm]
Constant Shear m=0,5	1.58	1290	0.41	33°	0.36
Constant Shear m=0,8	1.58	1440	0.46	31°	0.47
Shirakashi-Usui k=1,6	1.58	1300	0.40	34°	0.38
Sticking-Sliding μ=0,5	1.58	1305	0.41	34°	0.39
Sticking-Sliding μ=1,0	1.58	1320	0.42	33°	0.41
Constant Shear m=0,5	2.08	1275	0.41	33°	0.37
Constant Shear m=0,8	2.08	1410	0.44	32°	0.45
Shirakashi-Usui k=1,6	2.08	1284	0.40	33°	0.38
Sticking-Sliding μ=0,5	2.08	1281	0.40	34°	0.38
Sticking-Sliding μ=1,0	2.08	1290	0.41	34°	0.39

Figures 4 and 5 allow comparison of the predicted cutting forces and the experimental measurements for the two analysed cutting speeds. As well Figures 6 and 7 report the predicted chip thickness and contact length values as the friction model varies, compared with the experimental data for the highest cutting speed.

On the basis of the results reported in Table 2 and Figures 4–7, it is possible to assess the following relevant conclusions:

– the friction models investigated and the assumed friction coefficients are generally able to provide a satisfactory simulation of the physical phenomenon. Taking into account in particular the cutting force, even if a significant scattering of the predicted values was found out, the error was always lower than 7%; similar conclusions may be drawn as far as the chip thickness and contact length are concerned;

– the latter results are probably the most relevant, since they permit a validation of the numerical models taking into account "local" values instead of "global" variables, such as force;

– if the hypothesised frictional conditions are heavier (i.e. if the assumed shear factor m, or the friction coefficient • in the sliding region increase), the chip contact length increases, the chip thickness increases and finally the shear plane angle decreases. Such results are in full agreement with the experimental observations reported by several researchers and are consistent with the most relevant theoretical models;

– finally even if the simple constant shear model does not take into account so far the experimental measurements of the stresses distributions on the rake face, it is able nevertheless to provide quite effective results through a proper calibration of the shear factor.

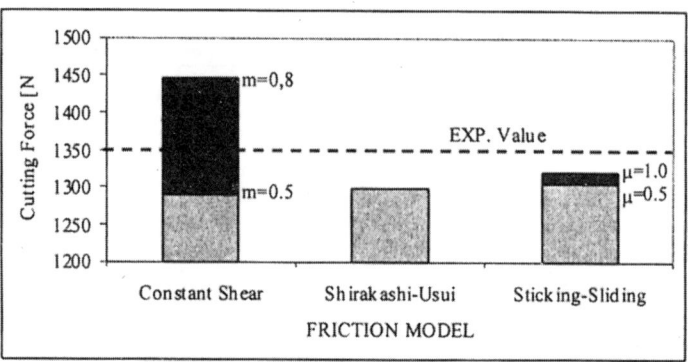

Figure 4. *Comparison of the numerical and experimental cutting force (cutting speed 1.58m/s)*

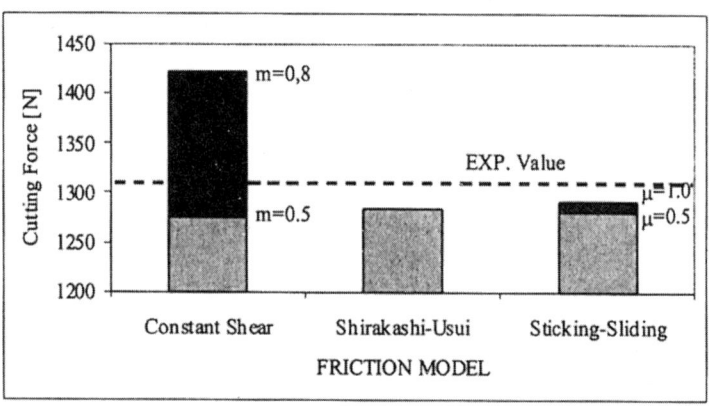

Figure 5. *Comparison of the numerical and experimental cutting force (cutting speed 2.08m/s)*

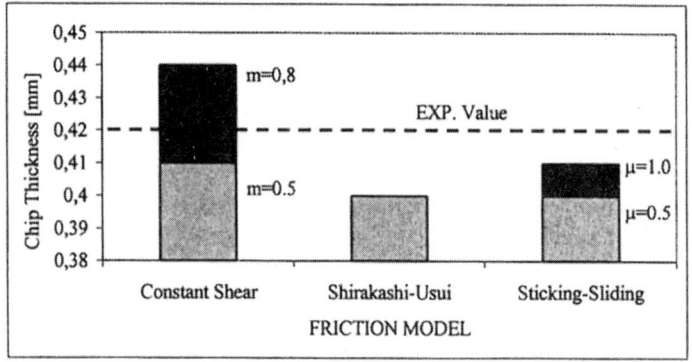

Figure 6. *Comparison of the numerical and experimental chip thickness (cutting speed 2.08m/s)*

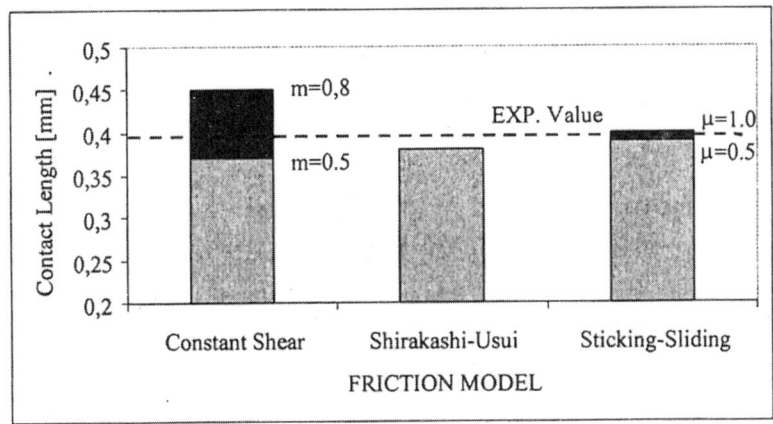

Figure 7. *Comparison of the numerical and the experimental contact length (cutting speed 2.08m/s)*

3. Orthogonal cutting with segmented type chip

It is well known that, depending on the cutting conditions, the chip shape may change from continuous to serrated and discontinuous. As a consequence, the capability to forecast the final chip shape would represent a very relevant tool in order to facilitate a better tool design and a more effective selection of the working parameters, improving the efficiency of the cutting operation and the final part quality [MAR 95] [KUM 97] [JOS 95] [LEE 51]. This fundamental task would be achieved without excessive experimental tests and within a reasonable simulation time.

The latter section of the present contribution is thus aimed to investigate the effect of friction model and friction value on the chip morphology prediction. In particular, two friction models were implemented, namely the Coulomb model and constant shear model, and the simulation results were analysed in terms of the predicted chip shape.

A customised release of DEFORM 2D was used to simulate orthogonal cutting with segmented chip formation [CER 97], [CER 98]. This is an implicit code suitable to calculate the state of stress, strain, strain rate and temperature inside the material during plastic deformation, but it did not include, in the basic release, the separation of the material. For this reason, in order to simulate the material breakage a new subroutine was linked to the original code. In particular material fracture was simulated by deleting those elements of the mesh for which the damage value is higher than an assigned critical value. Several damage criteria were tested by the authors (Cockroft & Latham, McClintock, Oyane, maximum effective strain [COC 66], [MCC 68], [ATH 97], [KLA 95]), but the one which seemed to be the most effective to simulate segmented chip formation was a damage criterion based on the maximum shear stress.

The maximum shear stress damage criterion is based on the assumption that the breakage of the material occurs in the primary deformation zone when the maximum shear stress is higher than a critical value τ_{lim}. A FORTRAN subroutine was implemented in the FEM code to calculate the maximum shear stress and to compare such value with the critical one: if τ_{max} is higher than the critical τ_{lim}, the element is deleted [CER 97], [CER 98]. To compensate the material loss related with the deleting of the elements, a smoothing subroutine was linked with the software. This subroutine smoothes the border of the chip, compensates the volume loss and facilitates the convergence of the simulation software [CER 99].

An orthogonal cutting process on low carbon free cutting steel (LCFCS) specimens using cemented carbide tools (P20 grade) was analysed. All over the simulations the cutting speed was fixed at 1.250 m/s, the depth of cut was 0.1 mm and the tool rake angle was -6°. The critical shear stress values was fixed at 510 MPa; this value was derived from the material breaking stress.

Figure 8 shows a typical shear stress distribution. The maximum shear stress is located in the primary shear zone, with the highest value very close to the free surface of the chip. Thus this is the area where initial fracture takes place and the chip begins to form.

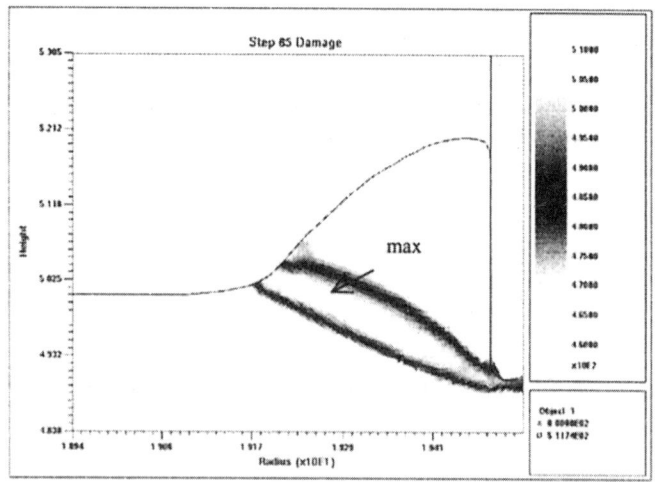

Figure 8. *Shear stress distribution and location of the maximum shear stress*

To test the sensitivity of friction modelling on simulation results two friction laws were considered, namely the constant shear model and the Coulomb law. In the former case a couple of friction factors were investigated, namely m = 0.5 and m = 0.82, while in the latter case the friction coefficient μ was a function of the normal stress on the rake face, according to [LUT 98]. Table 3 summarises the simulations carried out.

Table 3. *Simulations conducted*

Simulation	Friction model	Friction value
1.	Shear	0.82
2.	Shear	0.5
3.	Coulomb	Variable

Figures 9 and 10 show the segmented chip formation as the cutting tool advances (tool path 1 mm, and 2 mm) for the simulation n.1. Breakage is located in the primary deformation zone; it starts form the free surface of the chip and develops inside the chip itself. The chip morphology is serrated and the chip forms uniformly (the segments of the chip are regular and occur after the same tool path).

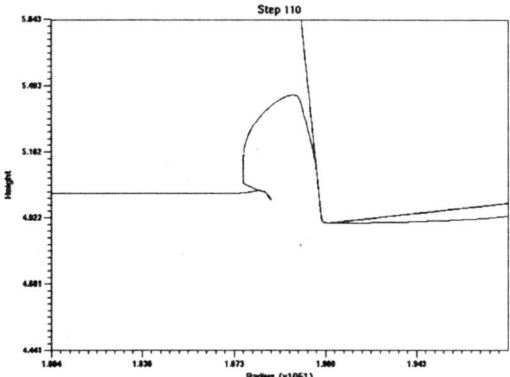

Figure 9. *Serrated chip: simulation set n.1, tool path 1mm*

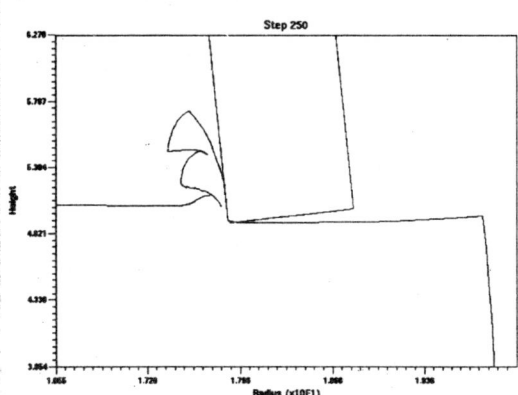

Figure 10. *Serrated chip: simulation set n.1, tool path 2mm*

Figures 11 and 12 show the chip morphologies obtained for the other investigated simulation sets.

The analysis of these results permits one to conclude that the predicted chip shape depends on the assumed friction model; in particular, for the same critical value of the shear stress, the following conclusions can be stated:

– a serrated chip is predicted, both utilizing the constant shear model with m = 0.82 and the Coulomb law with variable friction coefficient (Figures 10 and 12). In the latter case, the critical shear stress is reached before, and consequently for the same tool path more segments are observed in the chip;

– a continuous chip is predicted utilizing a friction factor m = 0.5, i.e. the material does not reach the critical shear stress and no elements are deleted (Figure 11).

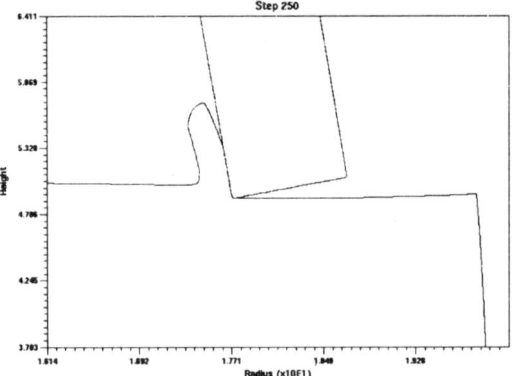

Figure 11. *Continuous chip: simulation set n.2, tool path 2mm*

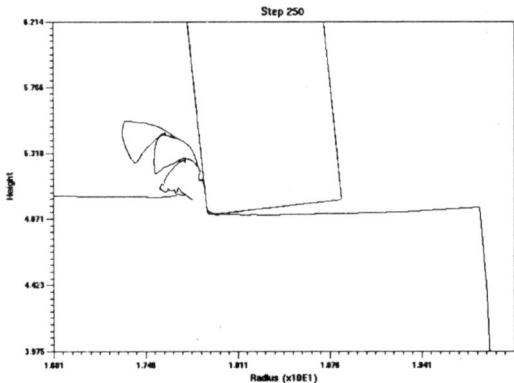

Figure 12. *Serrated chip: simulation set n.3, tool path 2mm*

In conclusion friction strongly affects the stress state and in particular the maximum shear stress value, determining a relevant variation so far as the predicted chip morphology is concerned. Increasing friction results in a segmented-type chip instead of a continuous one, while changing the friction model results in a more segmented chip with a lower chip spacing.

Actually a more extensive comparison with experimental data has to be carried out in the next future in order to evaluate the reliability of the model with different workpiece materials, cutting tool materials and geometries and cutting parameters. Furthermore, other damage shear plastic energy on the shear plane will be investigated. criteria will be tested. In particular, the effectiveness of a new criterion based on the maximum

Acknowledgements

This work has been made using MURST (Italian Ministry for University and Scientific Research) funds.

4. References

[STR 85] STRENKOWSKI J.S., CARROLL J.T., "A finite element model of orthogonal metal cutting", Journal. of Eng. for Ind., vol. 107, p. 349-354, 1985.

[STR 90] STRENKOWSKI J.S., MOON K.J., "Finite element prediction of chip geometry and tool/workpiece temperature distributions in orthogonal metal cutting", Journal. of Eng. for Ind., vol. 112, p. 313-318, 1990.

[SHI 95] SHIH A.J., "Finite element simulation of orthogonal metal cutting", Journal. of Eng. for Ind., vol. 117, p. 84- 93, 1995.

[MAR 95] MARUSICH T.D., ORTIZ M., "Finite element simulation of high-speed machining", Proc. of NUMIFORM'95 (1995), p. 101-107.

[VAZ 98] VAZ M. et al., "Finite element techniques applied to high-speed machining", Proc. of NUMIFORM'98 (1998), p. 973-978.

[SHI 83] SHIRAKASHI T. et al., "Flow stress of low carbon steel at high temperature and strain rate", Bulletin of JSPE, vol. 17, p. 167-172, 1983.

[LIN 92] LIN Z.C., LIN S.Y., "A coupled finite element model of thermo-plastic large deformation for orthogonal cutting", Journal. of Eng. Mat. and Tech., vol. 114, p. 218-226, 1992.

[ZOR 58] ZOREV N.N., "Results of work in the field of the mechanics of the metal cutting process", Proc. of the Conf. on Techniques of Engineering Manufacture (1958), p. 237-255.

[TRE 77] TRENT E.M., Metal Cutting, Butterworth, London, 1977.

[SHI 73] SHIRAKASHI T., USUI E., "Friction characteristics on tool face in metal machining", Journal JSPE, vol. 39, n° 9, p. 966-971, 1973.

[USU 82] USUI E., SHIRAKASHI T., "Mechanics of machining – From descriptive to predictive theory", On the art of cutting metals – 75 years later, ASME, vol. PED-7, p. 13-35, 1982.

[CER 97] CERETTI E. *et al.*, "Simulation of metal flow and fracture applications in orthogonal cutting, blanking and cold extrusion", Annals of CIRP, vol. 46/1, p. 187-190, 1997.

[CER 98] CERETTI E., "FEM simulations of segmented chip formation in orthogonal cutting: further improvements", Proc. of the First CIRP Int. Workshop on Modeling of Machining Operations (1998), p. 193-202.

[FIL 99] FILICE L., MICARI F., "Analysis of the relevance of some simulation issues on the effectiveness of orthogonal cutting numerical modeling", Proc. of the Second CIRP Int. Workshop on Modeling of Machining Operations (1999), p.270-282.

[LIN 97] LIN Z.C., LO S.P., "A study of the tool-chip interface contact problem under low cutting velocity with an elastic cutting tool", Journal. of Mat. Proc. Tech., vol.70, p.34-46, 1997.

[KUM 96] KUMAR S. *et al.*, 1996, Finite Element Simulation of Metal Cutting Processes: Determination of Material Properties and Effects of Tool Geometry on Chip Flow, Report No. ERC/NSM-D96-17, ERC for Net Shape Manufacturing, The Ohio State University.

[JOS 95] JOSHI V. S. *et al.*, Viscoplastic analysis of metal cutting by finite element method, Int. J. Mach. Tools Manufact., 34, 1994.

[LEE 51] LEE E.H. and SHAFFER B.W., 1951, The Theory of Plasticity applied to a Problem of Machining, Journal of App. Mech. Science, Vol. 7, p. 43.

[COC 66] COCKROFT M.G., LATHAM D.J., "A simple criterion of fracture for ductile metals", National Engineering Laboratory, Report 216, 1966.

[MCC 68] MC. CLINTOCK F.A., "A criterion for ductile fracture by the growth of holes", Journal. of Appl. Mech., vol. 35, p. 363-368, 1968.

[ATH 97] ATHAVALE S.M., STRENKOWSKI J.S., "Material damage-based model for predicting chip-breakability", Journal. of Eng. for Ind., vol. 119, p. 675-680, 1997.

[KLA 95] KLAMECKI B. E. E KIM S., On the plane stress to plane strain transition across the shear zone in metal cutting, J. Eng. Ind., 110, 1988.

[CER 99] CERETTI E., "Numerical study of segmented chip formation in orthogonal cutting", II CIRP International Workshop on Modeling on Machining Operations, Nantes, France - January 1999.

[LUT 98] LUTTERVELT C.A. *et al.* "The state-of-the-art of modeling in machining processes", Annals of CIRP, vol.47/2, p.587-626, 1998.

Chapter 6

Experimental Investigation and Prediction of Frictional Responses in the Orthogonal Cutting Process

Wit Grzesik
Dept of Manufacturing Engineering and Production Automation, Technical University of Opole, Poland

1. Introduction

The explanation and prediction of friction in the cutting process are still major problems that substantially limit the optimum shaping of ferrous and non-ferrous blanks. Nevertheless, friction-oriented cutting experiments, modelling and computer simulation of friction have resulted in the elaboration of practical methods leading to the reduction of friction and wear. Dauzenberg [DAU 99] has postulated that lower friction between the chip and the tool can be more effective than a harder tool material. All tribological activities associated with the implementation of low friction machining can be classified into the following subject groups:

− The reduction of the natural contact length (the friction zone) that leads to shorter contact time, lower contact temperature and lower friction between the rake face and the moving chip [SAD 93].

− The reduction of friction by applying coated tools layered with hard or hard and soft composite coatings. In the chip-tool tribo-system, coating plays the role of the "third" element, and from a tribological point of view it should result in the reduction of friction [BOW 73].

By deposition of the soft layer of MoS_2 on the hard layer of TiAlN, the coefficient of friction is drastically reduced to values of 0.05-0.15 [CSE 98]. In contrast, typical single and multilayer coatings operate at $\mu = 0.3$-0.4. In practice, in order to enhance the friction-lowering effect, the combined influence of the restricted contact and coatings is utilised for the design of moulded tool inserts. In fact, if the contact temperature is lowered, the hardness of the tool materials at a high cutting speed is still enough to remove the work material with high efficiency. Dry machining with a hot tool offers other real advantages. In particular, it can partly eliminate brittle mechanisms of tool fracture, such as micro-chipping and surface cracking, and increase the tool reliability [GRA 00].

− Minimum quantity lubrication. This technology is an answer to environmental pressure placed on manufacturers to eliminate the coolants typically used during machining. For example, in hard turning experiments with a bearing steel of 62-64 HRC and TiN coated tools, when an air-oil mist is supplied to the contact area, the tool life increases by about 30% in comparison with that of the CBN tool [KO 99].

− The use of vibrations, which cause the lubricant to penetrate easily into the contact zone.

− The use of workpiece materials with low friction additives, such as free-machining carbon and stainless steels containing sulphur, lead, lead and sulphur, sulphur and selenium [HAN 96].

Nowadays, it is generally accepted that improved tool life and increased productivity under dry and hard machining conditions can be achieved by applying complex multi-component and multilayer thin hard coatings deposited by means of CVD (ca. 43%) and PVD (ca. 10%) techniques [KLO 98, PRE 98, HUS 98]. Chemical vapour deposition (CVD) technology is currently used to produce multilayered coatings combining TiC, TiN, Ti(C,N) and Al_2O_3 films. On the other hand, physical vapour deposition (PVD) offers more wear-resistant Al_2O_3 coatings with controlled deposition of α- Al_2O_3 or κ- Al_2O_3 [SÖD 97] and a new generation of PVD-TiCN and PVD-TiAlN coatings that provides increased productivity in a broad range of machining operations and workpiece materials. According to world-class manufacturers of cutting tool materials [JIN 99, KLO 99] the future will be based on tough ceramics, diamond and PCBN-based composites in the body or coating of the tool.

Under these circumstances it is of primary importance to answer the question of how we can utilise the vital technological potential offered by coatings for controlling friction in the cutting process. It is obvious that additional research is necessary to clarify the tribology of coatings with application to the cutting process and to develop mechanistic and thermal models for the substrate/coating-chip contact.

2. Investigation of frictional behaviour of coated carbides

The determination of friction is one of the most essential problems emerging in the design of cutting tools and prediction of the tool life. The quantification of the frictional behaviour of tool coatings is usually based on friction-oriented cutting experiments and also on special friction testing that reproduces the contact conditions at the tool-chip interface.

In a direct cutting experiment, the friction and normal forces acting on the contact area are calculated in terms of the measured cutting and feed forces. This data is used to determine the coefficient of sliding friction and, for the given actual contact area, the contact stresses and the density of friction power (the frictional heat flux).

Unfortunately, we are not able to reproduce in full-scale the contact conditions in machining using conventional pin-on-disc testing because the wear mechanisms involved are not relevant to that observed in machining. In order to minimise this discrepancy, modified pin-on-disc [OLS 89, MEI 00], pin-on ring [HED 91] and ball-on-disc [WIK 99] test devices have been developed in order to perform sliding wear tests that simulate dry machining when using coated cutting tool materials. Hedenqvist and Olsson [HED 91] have shown that the coating tested and the substrate should be considered as one element termed *a coating/substrate system*. Recently, Meiller et al. [MEI 00] have carried out extensive friction experiments using a special test device including the thermal and mechanical outputs of the plain-plain contact tribo-system. In this later and other studies [GRZ 98, GRZ 00] the vital role of the frictional heat flux (friction mechanical dissipation) in the frictional behaviour of the work (chip)-coating/substrate system was confirmed.

Most of the prior research has shown [GRZ 99a, GRZ 98, SAD 93] that coatings influence the chip-formation mechanism and the tribological interaction at the chip-tool contact area. It was found [GRZ 99b] that the optimum choice of the thermal properties of the coating components and the coating structure leads to a reduction of friction between the chip and the rake. As a result, a substantial decrease in the mechanical and thermal loads acting in the vicinity of the cutting edge was observed. M'Saoubi *et al.* [MSA 98] have investigated the distribution of temperatures in the cutting zone using the CCD- infrared technique. They concluded, based on the experimental thermal maps predicted for uncoated and coated tools, that a top layer of TiN with lower friction properties reduces the temperature near the cutting edge. Moreover, they confirm the generation, under specific contact conditions, of a thermal barrier effect provided by a thin ceramic intermediate layer, which was earlier suggested in [GRZ 98].

To achieve a more realistic view into tribo-contact behaviour, new techniques for contact image processing should be developed. As a consequence, such important geometrical outputs as the contact length and the contact area can be dimensioned more accurately.

In this study, a set of tribo-contact characteristics involving the friction force, the frictional heat flux, mechanical contact stresses and the coefficient of friction in terms of kinematic and geometrical inputs (the feed rate, the cutting speed and the interface control factor) were identified.

3. Experimentation

3.1. *Experimental procedure*

The purpose of these experiments was to obtain data for predicting the friction behaviour and corresponding friction heat division for orthogonal cutting when using coated tools. The experimental methods employed in this study are similar to those reported by the author in previous research concerning the influence of coatings on the cutting process [GRZ 99b, GRZ 98].

The experimental program consisted of several series of turning tests, which were carried out on a precision lathe equipped with force and temperature measuring systems (Figure 1).

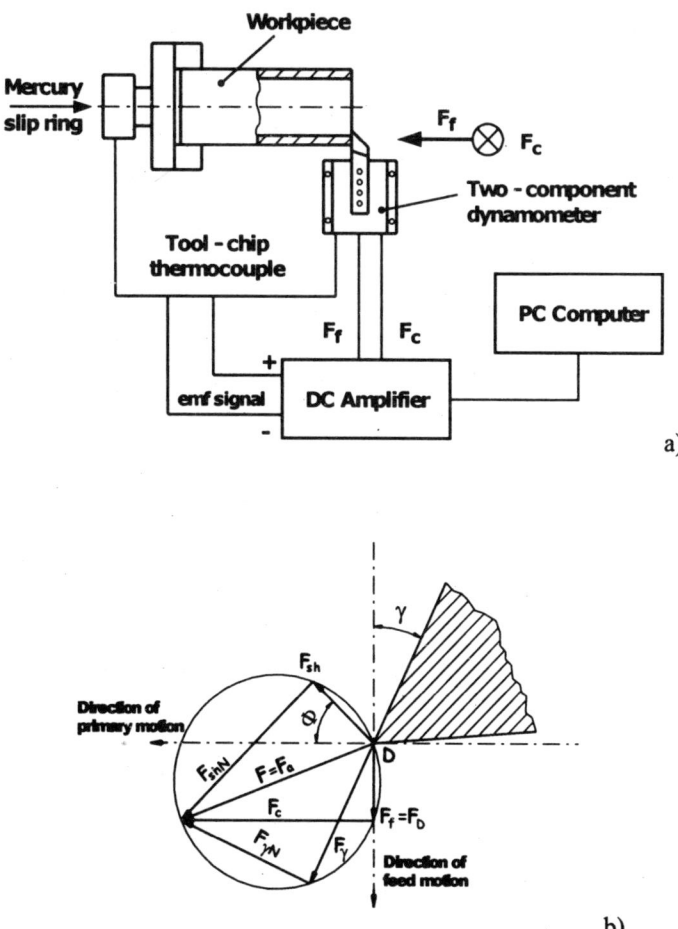

Figure 1. *Scheme of the experimental set-up (a) and force resolution in orthogonal cutting after ISO 3002/4 (b)*

In this investigation the following conditions were used:

– Workpiece: thin-walled tubes of AISI 1045 carbon and AISI 304 austenitic stainless steels, 2 mm thickness, the outer diameter of the tube was varied to obtained ca. 20% increment of the cutting speed. In cuts with varying feed rate an outer diameter of 80 mm was kept.

– Tool materials: uncoated tungsten carbide P20, single-layer (TiC), two-layer (TiC/TiN), three-layer (TiC/ Al_2O_3/TiN) and four-layer (TiC/Ti(C,N)/Al_2O_3/TiN) coated inserts.

In Table 1 values of the thermal conductivity and thermal diffusivity for the tested materials at boundary temperatures, 300 K and 1000 K, are provided.

– Tool configuration: flat-faced rake, rake angle γ_o=-5^0.

– Cutting speed: v_c = 30–210 m/min.

– Feed rate: f = 0.08-0.28 mm rev^{-1}. For the cutting arrangement used the feed rate was equal to the undeformed chip thickness.

– Depth of cut: 2 mm in all cutting trials.

Table 1. *Selected physical properties of coating components and steels used*

Type of coating		Thermal conductivity k, W/(mK) k at 300K/k at 1000K
CVD-TiC CVD-TiN CVD-Al$_2$O$_3$		32.0/41.1 20.0/25.7 36.0/5.0
Steel grade	Thermal conductivity k, W/(mK) k at 300K/k at 1000K	Thermal diffusivity $\alpha \times 10^6$, m^2/s α at 300K/α at 1000K
AISI1045 carbon steel	45/28.6	12.95/3.75
AISI304 stainless steel	14.9/25.4	3.95/5.26

As a consequence, ten different tribo-pairs, two P20 carbide-on-steel and eight coating-on-steel pairs, were tested. Experiments were replicated three times for each set of cutting conditions in order to minimise measuring errors.

3.2. Measurements

All signals generated in the cutting zone and the effects of friction at the interface were measured using either *in-process* or *post-process* measuring techniques. The structure of the experimental set-up and the measuring circuits are shown in Figure 1a. In order to measure the cutting and feed forces, a 2D strain-gauge dynamometer fixed on the tool post of a lathe was used. The thermal output of the tribo-system, i.e. the thermal *emf* signal generated in a hot junction produced by the top surface of the cemented carbide insert or the top layer of the coating and the moving chip, was recorded using the tool-work thermocouple circuit. In both cases a DAQ system consisting of an amplifier, an A/D converter, a PC computer and data acquisition software was linked to the measuring circuits.

After cutting, wear patterns on the rake face, occurring as a result of attrition in a running-in period, were visualised, and in subsequent stages the raw colour image was processed in order to define the sharply outlined contours of the contact zone. Figure 2 shows the contact estimation procedure including the visualisation of the selected fragment of the coating surface, computer image processing and automatic dimensioning of the real contact area.

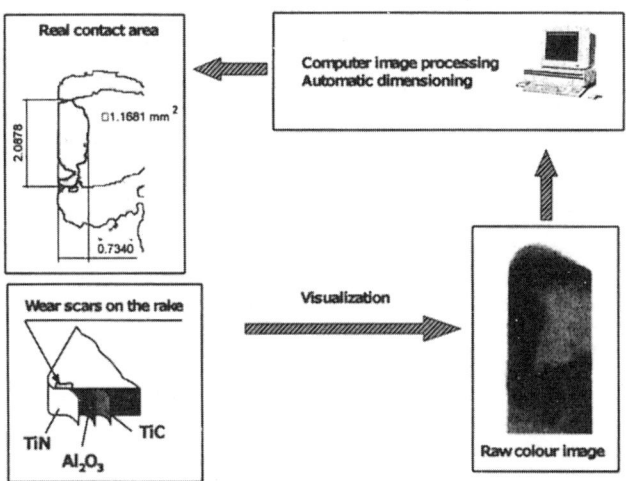

Figure 2. *A scheme of image processing system for dimensioning of the contact area*

3.3. Calculations

In the case of a simplified model for orthogonal cutting (Figure 1b) the resultant cutting force can be resolved into components normal and parallel to the rake face using the so-called Merchant's circle [SHA 89]. They are defined as the force normal to the rake face ($F_{\gamma N}$) and the friction force (F_{γ}) respectively. For given forces and the estimated contact area A_c, both the shear (τ_f) and normal stresses (σ_f) acting at the tool-chip interface and the mean coefficient of friction (μ_{γ}) were calculated. The frictional heat flux (q_f) is computed as the ratio of the friction power to the real contact area. Formulae used for calculating the selected experimental responses are specified in Table 2.

Table 2. *Specification of calculated quantities*

Number	Calculated quantity	Equation	References
1	Friction force	$F_\gamma = F_c \sin\gamma_o + F_f \cos\gamma_o$	[SHA 89]
2	Normal load	$F_{\gamma N} = F_c \cos\gamma_o - F_f \sin\gamma_o$	[SHA 89]
3	Interface control factor	$K_{int} = l_{nc}/h_D$	[GRZ 97]
4	Normal contact stress	$\sigma_f = \dfrac{F_{\gamma N}}{A_c}$	[SHA 89]
5	Shear contact stress	$\tau_f = \dfrac{F_\gamma}{A_c}$	[SHA 89]
6	Mean coefficient of sliding friction	$\mu_\gamma = \dfrac{\tau_f}{\sigma_f} = \dfrac{F_\gamma}{F_{\gamma N}}$	[SHA 89]
7	Frictional heat flux	$q_f = \dfrac{F_\gamma v_{ch}}{A_c} = \dfrac{F_\gamma v_c}{\lambda_h A_c}$	[GRZ 99b]

Nomenclature: F_c - cutting force, F_f - feed force, γ_0 - orthogonal tool rake angle, l_{nc} - natural contact length, h_D - undeformed chip thickness, A_c - actual area of contact, v_c - cutting velocity, v_{ch} - chip velocity, λ_h - chip thickness compression ratio

4. Results and discussion

4.1. *Forces and stresses at the interface*

Based on the experimental data, the tangential and normal forces exerted at the contact part of the rake face and, subsequently, the corresponding stresses acting on this region can be determined. A visual difference can be observed between the friction forces at the tool-chip contact obtained for all selected coatings, as shown in Figures 3 and 4. The difference related to four types of coatings becomes pronounced for a TiC/TiN coating and it is found, at higher feed rates, to be independent of steel grade. This implies that in this case more energy is needed to overcome friction and for the shearing. In contrast, a minimum friction force corresponds to both workpieces sliding over four-layer coating. This evidence is supported in Figure 5 (graph 2a and 2b), where the tangential force increases more intensively with a slow increase in the contact length. Moreover, Figure 5 demonstrates clearly that there exists a distinct boundary between the two steels cut. This indicates that there are different tribological conditions generated for both tool and work materials coupled, but the extreme difference occurs for uncoated carbide tools.

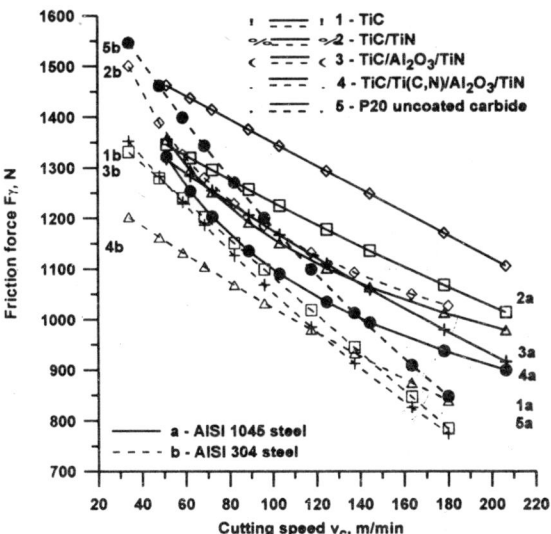

Figure 3. *Effect of cutting speed on the friction force*

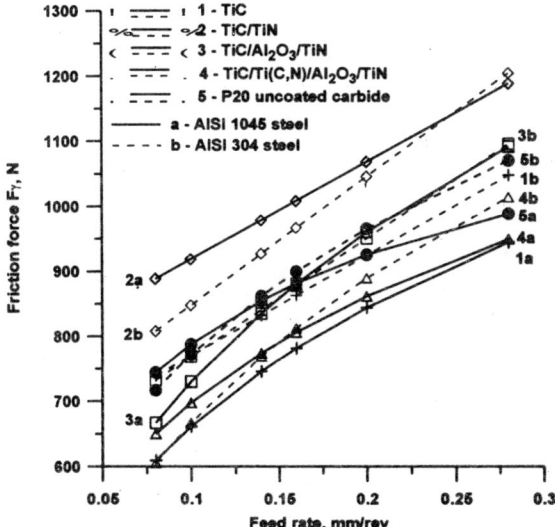

Figure 4. *Effect of feed rate on the friction force*

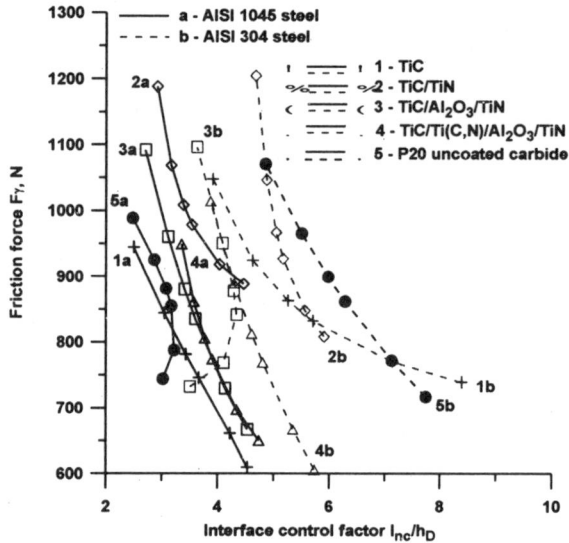

Figure 5. *Effect of the interface control factor on the friction force*

Figure 6 shows that, in general, the contact area increases with an increase in the feed rate, and this area estimated for an AISI 304 stainless steel is distinctly higher than that for an AISI 1045 carbon steel. The influence of coating leads, especially in the case of the four-layer coating represented by graphs 4a and 4b, to the reduction (AISI 304 steel) or enlargement (AISI 1045 steel) of the contact area.

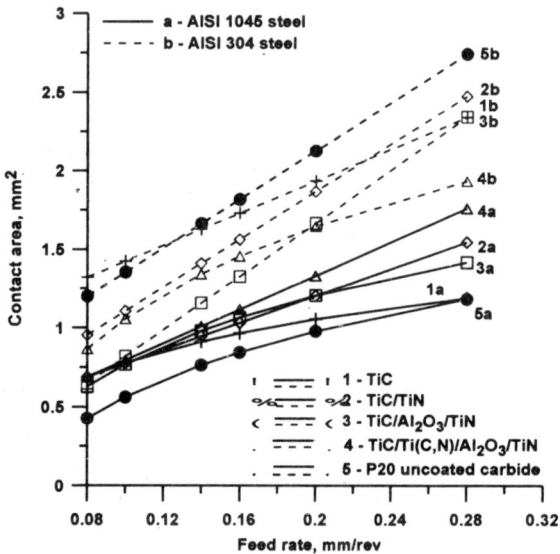

Figure 6. *Effect of feed rate on the contact area*

The arrangement of the shear contact stress τ_f versus normal contact stress σ_f, presented in Figure 7, indicates that their values depend strongly on the steel grade and the type of coating employed. Some combinations, as for example, three-layer $TiC/Al_2O_3/TiN$ coating against AISI 304 steel (graph 3b) may lead to a significant increase in the values of the shear and normal stresses. In contrast, four-layer coating containing Al_2O_3 ceramic film is likely to reduce the contact loads, as in machining of carbon steel (graph 4a). The same effect occurs for the AISI 304-TiC pair represented by graph 1b. In this investigation the normal stresses determined for coated flat-faced inserts were varied from 1270 MPa to 2050 MPa for medium carbon steel and from 720 MPa to 1440 MPa for stainless steel.

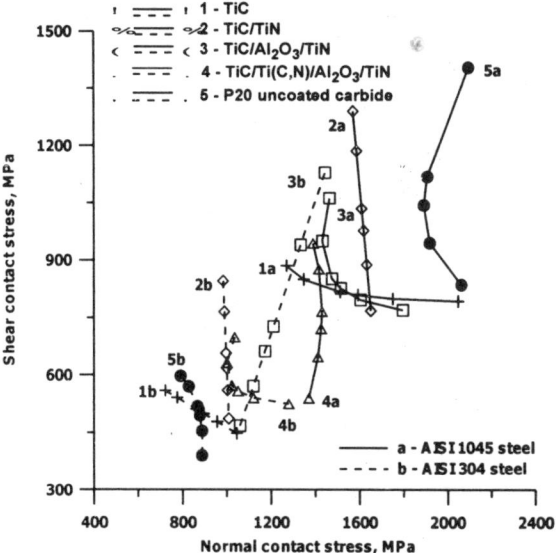

Figure 7. *Shear contact stress versus normal pressure*

In general, all tribo-contacts including stainless steel are less sensitive to variations in the shear stress as the normal pressure working on the interface increases. It is evident, from the above consideration, that knowledge about the thermal and tribological properties of each coating-work material pair creating a very specific tribo-system and possible ways of modifying them is crucial for lowering the tool-chip interface loading and, as a final result, for an increase in tool-life.

4.2. Friction and frictional heat division

In this tribological study, friction action was expressed in terms of the classical coefficient of friction μ_γ and the friction heat partition q_f flowing to one of the components of the tribo-contact pair. Figures 8 to 11 illustrate in a comprehensive way the final effect of three different process variables represented by the cutting speed (Figure 8), the normal load (Figure 10), and the interface control factor (Figure 11), and additionally by the contact temperature resulting from variations of the feed rate (Figure 9).

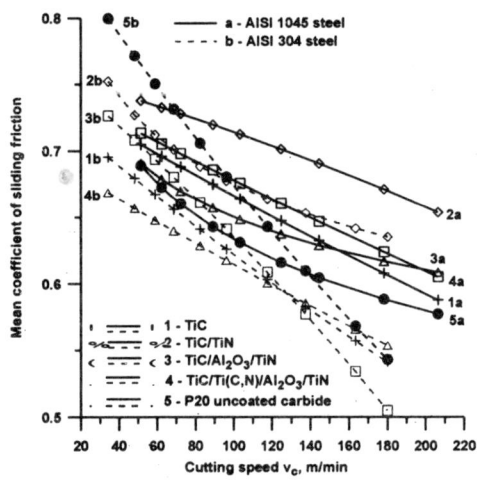

Figure 8. *The mean coefficient of friction versus cutting speed*

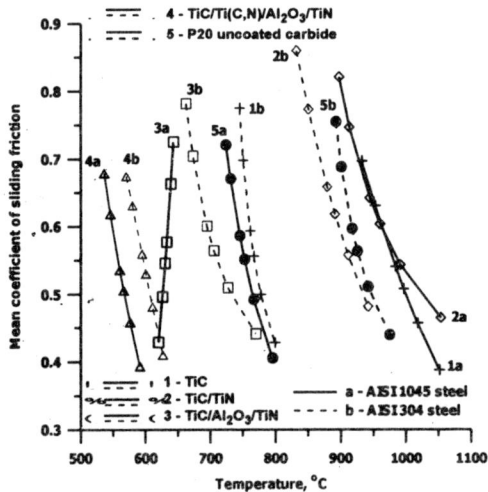

Figure 9. *The mean coefficient of friction versus contact temperature*

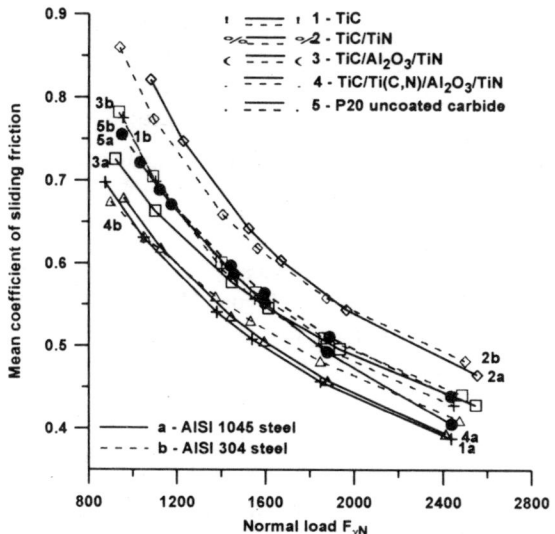

Figure 10. *The mean coefficient of friction versus normal load*

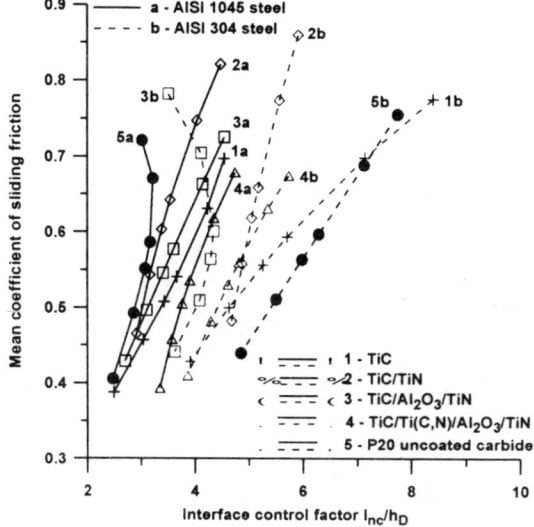

Figure 11. *The mean coefficient of friction versus the interface control factor*

Figure 10 shows a quite uniform reduction of μ_γ caused by an increase in the normal load. In this case the normal force was varied from 0.90 kN to 2.5 kN for the two steels used. Under the thermal conditions generated, when varying feed rate (Figure 9), the values of μ_γ change in a similar range from 0.4 to 0.8 for all coatings used (positive slope is only observed for graph 3a that represents AISI 1045-TiC/Al$_2$O$_3$/TiN pair), and for uncoated carbide tools. It is very important to note that while the coefficient of friction varies, the contact temperature measured for these pairs remains practically constant - about 600°C for the AISI 1045-TiC/ Al$_2$O$_3$/TiN pair and about 800°C for the AISI 304-TiC pair. It may thus be seen that the thermal softening effect due to the influence of the feed rate on the temperature rise is rather small.

The pronounced softening effect is regularly observed as a result of the increase in the cutting speed, as seen in Figure 8. Figure 11 depicts the obvious dependence of friction on the contact length. It appears from the graphs presented in this figure that stainless steel is more sensitive to changes in the contact length caused by coatings than a carbon steel.

Based on Figures 12 to 14 some tribo-contacts corresponding to the extreme intensity of the frictional heat flux were selected. Figures 12 and 14 confirm that both the cutting speed and feed rate affect the rate of frictional power dissipation at the interface. It should be pointed out that the cutting speed influences the frictional power directly and in the case of the feed rate this is largely due to the influence on the contact area.

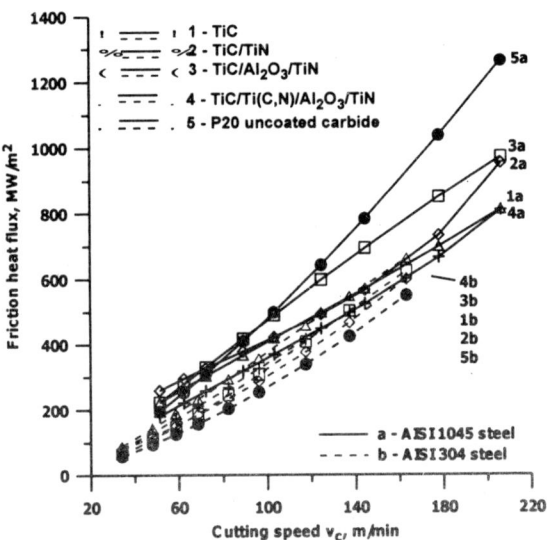

Figure 12. *Effect of cutting speed on frictional heat flux*

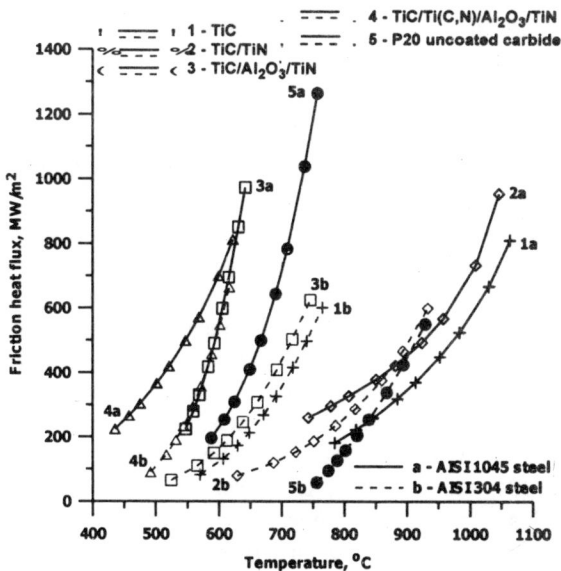

Figure 13. *Effect of the contact temperature on frictional heat flux*

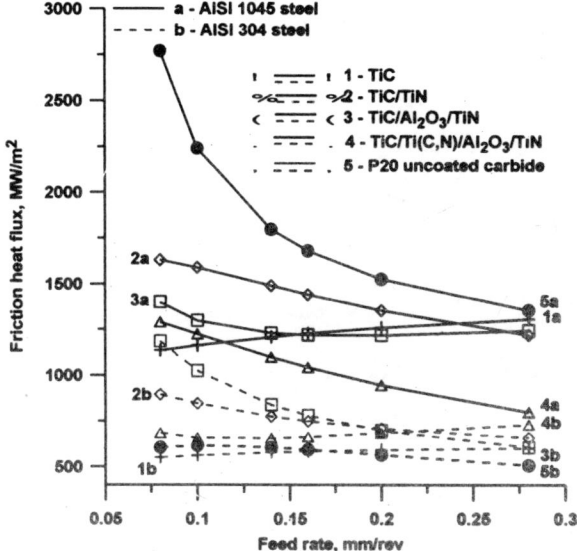

Figure 14. *Effect of feed rate on frictional heat flux*

The postulate that this parameter is significantly affected by changes in the contact temperature was examined in Figure 13. It was documented experimentally that for AISI 304-TiC/Ii(C,N)/Al₂0₃/TiN pairs (graph 4b) the frictional heat flux of

$660 \ MW/m^2$ occurs at a substantially lower temperature (ca. 600°C) than for uncoated carbide tools.

At a lower temperature, carbon steel has a thermal conductivity higher than stainless steel and can quickly dissipate the heat generated by friction. This is clearly explained by graph 3a in Figure 13, corresponding to three-layer coatings with an intermediate ceramic Al_2O_3 layer for which q_f approaches the value of $1000 \ MW/m^2$ at 650°C. Thus, the thermal barrier effect can occur due to a substantial difference between the thermal properties of the coupled materials at high contact temperature (see data in Table 1). As a result, much of the generated heat is transferred to the material with higher thermal conductivity, i.e. to the work material or the top layer of coating. It is demonstrated in Figure 14 that the selection of adequate coating is much more important when machining using smaller and moderate feed rates. A model compiled for the distribution of mechanical and thermal loads at the tool-chip interface for CVD-TiC/Ti(C,N)/Al_2O_3/TiN coating coupled with AISI 1045 carbon steel and AISI 304 stainless steel is shown in Figure 15a and Figure 15b, respectively. In this comparison, the interface conditions were predicted keeping a constant contact temperature of approximately 600°C.

a)

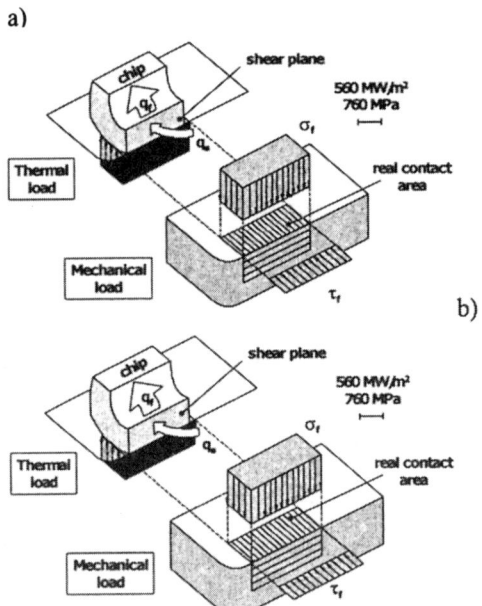

b)

Figure 15. *A model for the distribution of mechanical and thermal loads at the tool-chip interface*
a)AISI 1045–TiC/Ti(C,N)/Al_2O_3/TiN, f=0.16 mm/rev, v_c=180 m/min
(A_c=1.56, σ_f=1050 MPa, τ_f=650 MPa, q_f=700 MW/m^2, μ=0.620);
b)AISI 304–TiC/Ti(C,N)/Al_2O_3/TiN, f=0.16 mm/rev, v_c=135 m/min
(A_c=1.53 mm^2, σ_f=1045 MPa, τ_f=615 MPa, q_f=540 MW/m^2, μ=0.590)

5. Concluding remarks

From the experimental results, the following practical conclusions can be drawn.

– It was proven that hard coatings are responsible for the changes in friction at the interface with a sliding (cutting) speed commonly used in many machining operations.

– The friction force decreases substantially with the rise in the cutting speed and under the contact conditions for which the interface control factor approaches maximum value.

– Wide variability is seen in reported values of the friction coefficient for the thin coatings tested on sintered tungsten carbide sliding over the surface of a steel workpiece. The values of μ obtained for coating-on-metal contacts lie in the range from 0.4 to 0.8. These are similar to the values valid for ceramic-ceramic contacts and metallic couples sliding in air in the presence of intact oxide films [HUT 92].

– There is considerable interest in the use of Al_2O_3 ceramic as the intermediate layer in multilayered coatings to produce the thermal barrier effect. This is manifested by observation of the four-layer coating for which frictional heat is dissipated effectively at a temperature sufficiently lower than for other tribo-pairs tested.

– Both feed rate and cutting speed affect the rate of frictional power dissipation at the interface. In particular, the feed rate was found to act indirectly through the influence on the contact area.

From these conclusions, the friction magnitude can be predicted more reliably for typical CVD-coatings coupled with carbon and austenitic stainless steels.

The data obtained can be used to help choose from among the wide variety of coatings available today, attain dry cutting at higher speeds and achieve longer tool-life.

6. References

[GRA 00] GRAHAM D., "Dry machining. Going dry", *Manufacturing Engineering*, vol. 124, no. 1, 2000, p.72-78.

[GRZ 00] GRZESIK W., "The influence of thin hard coatings on frictional behaviour in the orthogonal cutting process", *Tribology International*, vol. 33, 2000, p.131-140.

[MEI 00] MEILLER M., LEBRUN J.L., TOURATIER M. *et al.*, "Friction Law for Tool/ Workpiece Area in Dry Machining", Proceedings of the International Workshop on Friction and Flow Stress in Cutting and Forming, 2000, p.101-109.

[DAU 99] DAUTZENBERG J.H., TAMINIAU D.A. *et al.*, "The workpiece material in machining", *Int. J. Adv. Manuf. Technol.*, vol. 15, 1999, p.383-386.

[KO 99] KO T.J. *et al.*, "Air-oil cooling method for turning of hardened material", *Int. J. Adv. Manuf. Technol.*, vol. 15, 1999, p.470-477.

[GRZ 99a] GRZESIK W., "Experimental investigation of the cutting temperature when turning with coated indexable inserts", *Int. J. Mach. Tools Manuf.*, vol. 39, 1999, p. 355-369.

[GRZ 99b] GRZESIK, W., "An integrated approach to evaluating the tribo-contact for coated cutting inserts", *Proceedings of the CIRP Int. Workshop on Modelling of Machining Operations*, CD-ROM version, Nantes, 1999.

[JIN 99] JINDAL, P.C., SANTHANAM, A.T., SCHLEINKOFER, U., "PVD coatings for turning", *Cutting Tool Engineering*, vol. 51, 1999, p. 42-52.

[KLO 99] KLOCKE, F., KRIEG, T., "Coated tools for metal cutting-Features and applications", *Annals of the CIRP*, vol. 48/2, 1999, p. 1-11.

[WIK 99] WIKLUND, U., WÄNSTRAND, O., LARSSON, M., HOGMARK, S., "Evaluation of new multilayered physical vapour deposition coatings in sliding contact", *Wear*, vol.236, 1999, p. 88-95.

[CSE 98] CSELLE T., Carbide drills: at the peak of development?, Gühring OHG, 3rd ed.,1998.

[HUS 98] HUSTON M.F. *et al.*, "Cutting materials, tools, market trends in USA", *VDI Berichte*, n 1399, 1998, p.21-54.

[GRZ 98] GRZESIK, W., "The role of coatings in controlling the cutting process when turning with coated indexable inserts", *J. Mat. Proc. Technol.*, vol. 79, 1998, p.133-143.

[KLO 98] KLOCKE, F., KRIEG, T., GERSCHWILER, K. *et al.*, "Improved cutting processes with adapted coating systems", *Annals of the CIRP*, vol. 47/1, 1998, p. 65-68.

[PRE 98] PRENGEL, H.G., PFOUTS, W.R., SANTHANAM, A.T., "State of the art in hard coatings for carbide cutting tools", *Surface Coat. Technol.*, vol. 102, 1998, p. 183-190.

[MSA 98] M'SAOUBI R., LEBRUN J.L., CHANGEUX B., "A new method for cutting tool temperature measurement using CCD infrared technique: influence of tool and coating", *Machining Science and Technology*, vol.2, n 2, 1998, p. 369-382.

[SÖD 97] SÖDERBERG S., "Advances in metal cutting tool materials", *Scandinavian Journal of Metallurgy*, vol. 26, 1997, p. 65-70.

[GRZ 97] GRZESIK W., KWIATKOWSKA E., "An energy approach to chip-breaking when machining with grooved tool inserts", *Int. J. Mach. Tools Manufact.*, vol.37, n 5,1997, p. 569-577.

[HAN 96] A practical handbook. Modern metal cutting, Sandvik Coromant, 1996.

[SAD 93] SADIK, I. M., LINDSTRÖM, B., "The role of tool-chip contact length in metal cutting", *J. Mat. Proc. Technol.*, vol. 37, 1993, p.613-627.

[HUT 92] HUTCHINGS I.M., Tribology. Friction and wear of engineering materials. Edward Arnold, 1992.

[HED 91] HEDENQVIST, P., OLSSON, M., "Sliding wear testing of coated cutting tool materials", *Tribology International*, vol.24, n 3, 1991, p. 143-150.

[OLS 89] OLSSON, M., SÖDERBERG, S., JACOBSON, S., HOGMARK, S., "Simulation of cutting tool wear by a modified pin-on-disc test", *Int. J. Mach. Tools. Manuf.*, vol. 29, n 3, 1989, p. 377-390.

[SHA 89] SHAW, M. C., *Metal Cutting Principles*, Clarendon Press, Oxford, 1989.

[BOW 73] BOWDEN F. P., TABOR D., Friction. An introduction to tribology. Anchor Press/Doubleday, 1973.

Appendix 1. *Selected values of experimental results for AISI 1045 carbon steel. Cutting conditions: feed rate f = 0.16 mm/rev, depth of cut a_p = 2 mm, rake angle γ_0 = -5⁰*

	Cutting speed in m/min	Cutting temperature in ⁰C	Coefficient of friction	Heat flux in MW/m²	Friction force in N
P20 uncoated carbide	51.37	587.62	0.689	195.34	1918.06
	62.34	608.49	0.672	253.39	1863.12
	72.24	624.92	0.660	308.91	1822.38
	89.06	649.09	0.643	409.25	1766.11
	103.2	666.70	0.631	498.84	1727.58
	124.69	690.07	0.616	643.13	1679.39
	144.48	708.88	0.604	783.79	1642.83
	178.13	736.55	0.588	1038.13	1592.28
	206.4	756.72	0.577	1265.02	1557.68
TiC	51.37	785.01	0.704	180.72	1868.24
	62.34	818.73	0.696	221.28	1844.29
	72.24	845.40	0.688	257.51	1822.96
	89.06	884.80	0.674	318.54	1787.30
	103.2	913.66	0.663	369.73	1757.90
	124.69	952.12	0.647	448.62	1714.19
	144.48	983.21	0.632	524.23	1674.95
	178.13	1029.19	0.607	666.50	1610.38
	206.4	1062.84	0.587	810.72	1558.16
TiC/TiN	51.37	742.48	0.737	259.58	1982.12
	62.34	778.79	0.732	296.08	1961.55
	72.24	807.61	0.728	327.27	1942.99
	89.06	850.35	0.719	378.71	1911.46
	103.2	881.76	0.712	422.30	1884.94
	124.69	923.75	0.701	492.94	1844.65
	144.48	957.82	0.690	567.27	1807.55
	178.13	1008.37	0.671	733.91	1744.45
	206.4	1045.51	0.653	954.60	1691.45
TiC/Al₂O₃/TiN	51.37	547.00	0.713	222.88	1886.20
	62.34	559.15	0.705	279.01	1870.34
	72.24	568.62	0.698	330.12	1856.16
	89.06	582.39	0.685	417.08	1832.34
	103.2	592.31	0.675	489.64	1812.57
	124.69	605.33	0.660	597.92	1782.98
	144.48	615.71	0.646	694.64	1756.20
	178.13	630.81	0.623	850.91	1711.68
	206.4	641.69	0.605	973.12	1675.24
TiC/Ti(C,N)/Al₂O₃/TiN	51.37	435.80	0.690	223.04	1962.48
	62.34	457.95	0.678	265.70	1908.63
	72.24	475.56	0.669	303.48	1868.64
	89.06	501.75	0.657	366.55	1813.31
	103.2	521.05	0.648	418.90	1775.37
	124.69	546.91	0.637	498.08	1727.85
	144.48	567.93	0.628	571.42	1691.74
	178.13	599.22	0.616	699.39	1641.74
	206.4	622.25	0.608	812.61	1607.47

Appendix 2. *Selected values of experimental results for AISI 304 stainless steel. Cutting conditions: feed rate f = 0.16 mm/rev, depth of cut a_p = 2 mm, rake angle γ_0 = -5°*

	Cutting speed in m/min	Cutting temperature in °C	Coefficient of friction	Heat flux in MW/m²	Friction force in N
P20 uncoated carbide	34.20	755.91	0.800	58.14	1933.41
	48.09	774.52	0.772	94.83	1893.79
	58.81	788.88	0.751	126.35	1863.87
	68.71	802.14	0.732	157.61	1836.74
	82.33	820.39	0.706	203.69	1800.18
	96.19	838.96	0.681	254.02	1763.85
	117.62	867.67	0.643	338.57	1709.38
	137.41	894.18	0.610	424.39	1660.82
	163.36	928.94	0.568	550.33	1599.57
TiC	34.20	569.86	0.696	81.90	1944.72
	48.09	606.66	0.679	131.43	1888.55
	58.81	629.77	0.667	172.57	1846.35
	68.71	648.37	0.656	212.07	1808.25
	82.33	670.83	0.641	267.93	1757.15
	96.19	690.91	0.626	325.73	1706.69
	117.62	717.94	0.603	415.38	1631.59
	137.41	739.70	0.583	497.01	1565.27
	163.36	764.82	0.557	600.59	1482.50
TiC/TiN	34.20	629.75	0.752	79.29	1994.70
	48.09	686.70	0.727	119.45	1910.35
	58.81	722.53	0.712	152.98	1862.30
	68.71	751.40	0.701	185.89	1826.00
	82.33	786.32	0.688	234.20	1784.74
	96.19	817.56	0.678	287.04	1750.02
	117.62	859.68	0.664	376.28	1706.16
	137.41	893.63	0.653	467.17	1673.03
	163.36	932.88	0.642	599.45	1636.96
TiC/Al₂O₃/TiN	34.20	524.26	0.726	65.46	1832.75
	48.09	565.52	0.708	109.97	1806.19
	58.81	591.56	0.694	148.97	1785.69
	68.71	612.60	0.680	187.95	1766.76
	82.33	638.09	0.661	245.42	1740.71
	96.19	660.94	0.641	307.46	1714.20
	117.62	691.82	0.609	408.00	1673.22
	137.41	716.76	0.577	502.78	1635.38
	163.36	745.65	0.534	624.27	1585.75
TiC/Ti(C,N)/Al₂O₃/TiN	34.20	492.00	0.669	90.19	1798.72
	48.09	516.86	0.657	144.35	1769.51
	58.81	532.11	0.648	189.38	1747.31
	68.71	544.22	0.640	232.69	1727.07
	82.33	558.64	0.629	294.08	1699.63
	96.19	571.35	0.618	357.81	1672.18
	117.62	588.21	0.601	457.08	1630.65
	137.41	601.58	0.586	547.91	1593.28
	163.36	616.82	0.566	663.75	1545.64

Chapter 7

Variable Tool-Chip Interfacial Friction in 2-D and 3-D Machining Operations

A.K. Balaji
Dept of Mechanical Engineering, University of Utah, USA

I.S. Jawahir
Center for Robotics and Manufacturing Systems, and Dept of Mechanical Engineering, University of Kentucky, USA

1. Introduction

Tool-chip interfacial friction plays a critical role in the mechanics of chip formation in machining and influences major machining performance measures, especially cutting forces, tool-wear and chip-form. Unfortunately, traditional Coulomb friction does not apply to a high strain-rate, high temperature plastic deformation process such as machining. Fundamental knowledge of the frictional conditions at the tool-chip interface is therefore necessary for predicting machining performance. Considerable early significant work on tool-chip interfacial friction has been reported by several researchers ([FIN 56], [ZOR 63], [WAL 64], [MAT 81]). Challen and Oxley proposed an asperity deformation model using slip-line fields to characterise the tool-chip friction in machining [CHA 84].

One of the major drawbacks in attaining complete predictive capability of models for machining has been the inability to characterise the actual tribological conditions at the tool-chip interface with the assumption of highly approximate boundary conditions. This paper especially highlights the fact that some of the traditional results from 2-D plane-strain conditions cannot be directly superimposed to solve complex 3-D machining problems due to the inherent variability of friction in 3-D processes. The paper discusses frictional effects in cyclic chip formation. A large variety of cases always involve curled chip formation; if the chips break at regular intervals into concise chip-forms, the process is very cyclic in repeatability. Non-uniquely repeatable, yet cyclic chip formation is also common in machining. Hence, the assumption of quasi-static, stable straight chip formation has to be critically questioned for a new impetus towards predictive modelling of the actual machining process. This contribution highlights research issues related to the role of variable friction in 2-D and 3-D machining with flat-faced and grooved tools.

2. Characterisation of tool-chip interfacial friction

Tool-chip interfacial friction in machining can be characterised by different methods and parameters. Establishing a unique standard for evaluating tool-chip interfacial friction has been a major roadblock for development of theories. Major parameters and methods commonly used in estimating friction in machining include:

– Friction angle (λ): This parameter is an offshoot of Merchant's pioneering work [MER 44] on shear plane solutions. However, it is restricted to 2-D analysis and is based on average friction which needs substantial input in terms of other variables such as chip thickness, shear angle, etc.

– Friction parameter (τ/k): This is defined as the ratio of the tool-chip interfacial shear stress to the shear flow stress of the nominal work material. This parameter is a more quantifiable parameter than the friction angle and has been used extensively in analytical models for the 2-D machining process with flat-faced [OXL

89], [DEW 79] and restricted contact tools [FAN 00]. However, once again its extension to 3-D machining remains a hurdle.

  - Friction force on the rakeface (R'): The resolved resultant force on the rakeface which is widely used in predictive models for chip side-flow [RED 99] is also an appropriate measure for friction at the rakeface in 3-D machining operations, typically in turning with a nose radiused tool.

  - Tool-chip contact length (h_n): This is a direct consequence of the tool-chip interfacial friction behaviour. However, experimental measurement of the tool-chip contact has been limited to post-mortem measurement of the wear traces. These measurements have been shown to vary with theoretical prediction since the wear traces measure the final and maximum tool-chip contact. Spaans [SPA 70] used ultrasonic methods to characterise the contact length. In-situ measurement of tool-chip contact has been carried out by some researchers [DOY 79], [MAD 97] at low cutting speeds with transparent cutting tools and soft metals. Their results are contradictory in terms of sticking and sliding friction shown by Zorev [ZOR 63].

  - Chip deformation parameters from metallographic analysis (see Figure 1):
  - Grain elongation angle (Γ): Brown [BRO 87] used the grain orientation angle in his chip strain analysis. In the current work, the angle of grain elongation (Γ) was measured in a stable region beyond the secondary deformation zone. However, in the secondary deformation zone (characterised by the plastic zone $\delta t_2''$ defined below) the grains are swept back at varying rates which can be defined by a variable angle Γ_2 which has the lower limit of $\Gamma_2 \approx 0°$ at plastic zone thickness of $\delta t_2'$ and maximum value of $\Gamma_2 = \Gamma$ at plastic zone thickness of $\delta t_2''$.

  - Plastic zone thickness ($\delta t_2'$ and $\delta t_2''$): The plastic deformation undergone at the secondary deformation zone can be evaluated by two parameters which are both measured with reference to the tool rakeface: (a) $\delta t_2'$ is the thickness of a thin layer at the tool-chip interface formed due to heavily retarded flow of chip material relative to the flow in the bulk of the chip (i.e. $\Gamma_2 \approx 0$) and (b) $\delta t_2''$ signifies the entire plastic zone: this includes the layer $\delta t_2'$ where $\Gamma_2 \approx 0$ and an overlying layer in which the grains are swept back with variable grain elongation (Γ_2) and severe sweeping back of the grains. Trent [TRE 88] attributed the layer $\delta t_2'$ to seizure of the chip material at the interface whereas Oxley [OXL 89] argued, on basis of slip-line field analysis, that this layer was due to the retarded flow at the interface.

3. Experimental work

  The experimental work spanned 2-D and 3-D machining operations on 1045 steel with both flat-faced and grooved tools. The grooved tools had standard chip-groove geometry with land length of 0.26 mm. For 2-D machining, the experimental setup involved machining of disks with 3 mm width of cut. For the 3-D machining experiments, nose radiused flat-faced tools (r_ε = 0.8 and 1.2 mm) were used to attain predominantly side-curled chips. However, when the same conditions were used in machining with a standard grooved tool, the chips were in mixed mode with predominant up-curl. A wide range of depths of cut (a = 0.4 - 2.5 mm) and feeds (f = 0.1- 0.4 mm/rev.) were used. Figure 2(a) and (b) shows the formation of side-curl dominated chips (chips break after contacting the rotating workpiece into flat C-shapes) and up-curl dominated chips (chips break after contacting the tool flank).

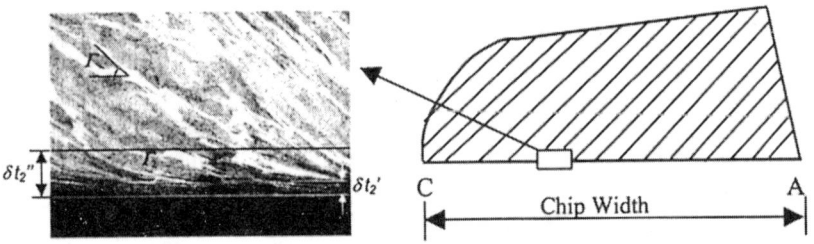

Figure 1. *Definition of plastic zone thickness and grain elongation angle for characterizing tool-chip interfacial friction across the chip width*

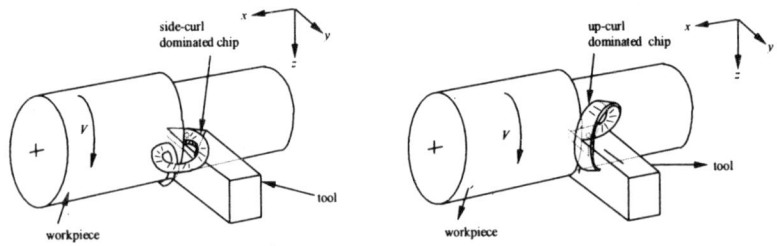

Figure 2. *Formation of 3-D curled chip: (a) side-curl dominated chip in machining with a flat-faced tool; (b) up-curl dominated chip in machining with a grooved tool*

4. Friction in 2-D machining with flat-faced and grooved tools

4.1. *2-D machining with flat-faced tools – quasi-static condition*

Most machining models have been limited to 2-D orthogonal machining with flat-faced tools, where plane-strain conditions and average friction are assumed and a straight continuous chip is usually expected. The concept of constant or average friction in 2-D machining is however, a marginal assumption even for the case of quasi-static machining not involving the effects of chip curl and breaking.

4.2. 2-D machining with flat-faced tools – cyclic chip formation

The machining process is an inherently dynamic process which features cyclic chip formation involving finite chip breaking cycles [GAN 98]. Fang and Jawahir [FAN 96] show that major frictional parameters vary within one chip breaking cycle (see Figure 3), thereby necessitating a non-unique appraisal of the problem of machining, even for the comparatively simple case of 2-D machining.

4.3. 2-D machining with grooved tools

Grooved tools are used to cyclically curl and break the chip for effective chip control applications. Jawahir and Zhang [JAW 95] have shown the effect of the groove backwall on the forces and bending moment developed within the chip. A new definition for evaluating tool-chip contact in grooved tools was presented by Balaji et al. [BAL 99a]. In this work, it was shown that the tool-chip contact in a grooved tool extends beyond the land into the secondary rakeface of the groove as well as at the backwall (see Figure 4).

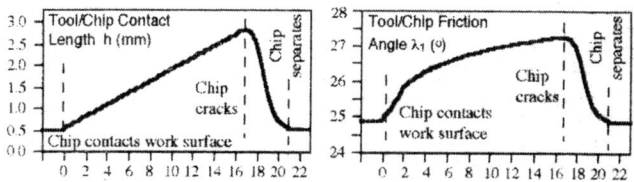

Distance traveled by the chip from the initial chip/work contact point (mm)

Figure 3. *Cyclic variations in tool-chip contact length and tool-chip friction angle in 2-D machining with a flat-faced tool [FAN 96].*

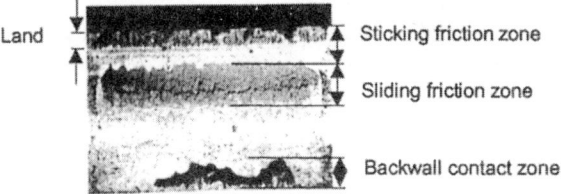

Figure 4. *SEM photograph showing sticking friction, sliding friction and backwall tool-chip contact zones (V = 100 m/min., f = 0.3 mm/rev., Work material = 1045 steel, Standard chip-groove) [BAL 99a].*

5. Tool-Chip Interfacial Friction in 3-D Machining

5.1. *Flat-faced tools*

In 3-D machining operations (e.g., turning) with a nose radiused flat-faced tool, variable tool-chip interfacial friction exists along the developed length of the cutting edge and on the tool rakeface. The reasons for the existence of variable friction are the non-linear effects induced by the nose radius: the variations in localised undeformed chip thickness (t_1), inclination angle (λ_s), rake angle (γ_n) and cutting edge angle (κ_r) (see Figure 5). This results in variable elemental friction force at every element of the undeformed area of cut. A further complication is induced by individual directions of elemental chip flow. However, in practicality, a chip is a single entity and the elements have to be in continuum. This results in varying frictional interactions at the tool-chip interface which dictate the chip curling pattern. Figures 6(a) and (b) show the variable tool-chip contact in turning with a nose radius tool. It is seen that, at a low feed, the tool-chip contact pattern is an approximate mirror image of the undeformed area of cut. At larger feeds, the tool-chip contact pattern shows a different pattern due to the strong interactions from different elemental chip-flow directions.

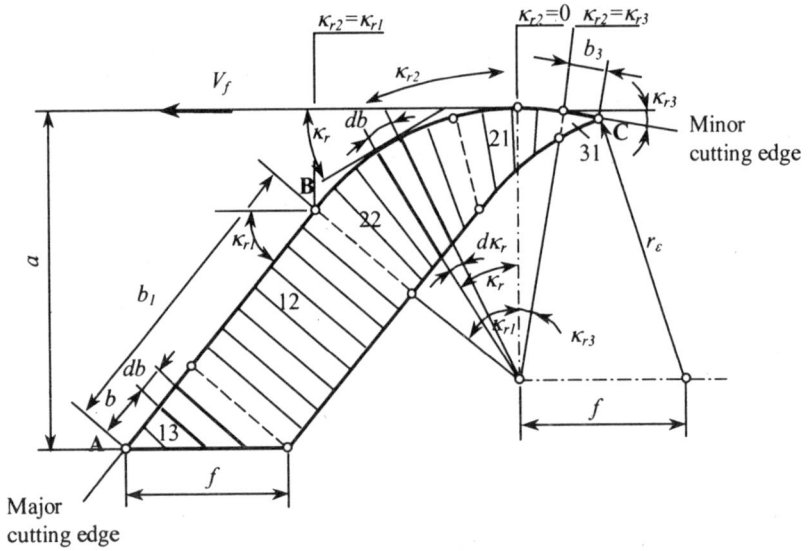

Figure 5. *The undeformed cut area and associated geometric parameters [RED 99].*

(a) f = 0.1 mm/rev. (b) f = 0.4 mm/rev.

Figure 6. *SEM photograph showing varying tool-chip contact on the rakeface* (Nominal tool geometry: inclination angle = -5 deg., rake angle = -5 deg., side-cutting edge angle = 90 deg., a = 2.5 mm, r_ε = 0.8 mm, V = 200 m/min., Work material: AISI 1045 Steel) *[BAL 99b].*

Chips produced in 2-D machining have an approximate rectangular transverse cross-section. In bar turning operations with a nose radius tool with dominant side-curled chip formation, the chip transverse cross-section (i.e., across the chip width) is not rectangular in shape [BAL 99b]. Depending on the depth of cut–nose radius ratio, the chip sections range from trapezoidal to triangular shape. Figures 6 (a) and (b) show the transverse chip cross-section for depths of cut of 2.5 mm and 1.2 mm respectively. This variation in chip thickness reflects the existence of variable friction at the tool-chip interface. The predictive model for chip side-flow in nose radiused flat-faced tools by Redetzky et al [RED 99] uses an elemental approach. Figure 5 has shown the basic geometry of the undeformed area of cut and the division into small elements along the developed length of the cutting edge. This model can be used to predict the elemental friction force components on the rakeface (dF_x' and dF_y') in the mutually perpendicular X and Y directions as well as the resultant elemental friction force on the rakeface (dR') as follows:

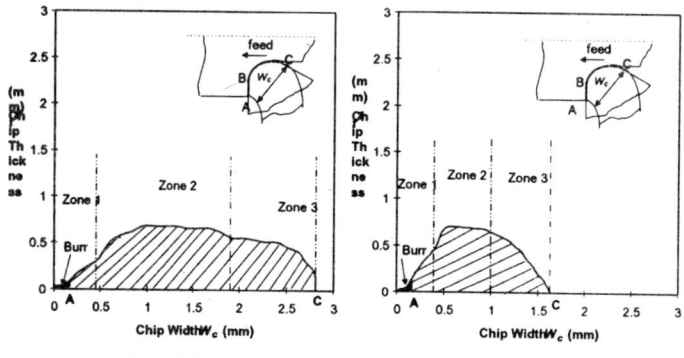

(a) a = 2.5 mm   (b) a = 1.2 mm

Figure 7. *Variation of chip thickness across the chip width* (Nominal tool geometry: inclination angle = -5 deg., rake angle = -5 deg., side -cutting edge angle = 90 deg.; f = 0.3 mm/rev., r_ε = 0.8 mm, V = 200 m/min., Work material: AISI 1045 Steel) *[BAL 99b].*

$dF_y' = (-dF_{Ax} \cos \kappa_{rl} + dF_{Ay} \sin \kappa_{rl}) \cos \lambda_{sl} + dF_{Az} \sin \lambda_{sl}$ [1]

$dF_x' = (dF_{Ax} \sin \kappa_{rl} + dF_{Ay} \cos \kappa_{rl}) \cos \gamma_{nl} + (dF_{Az} \cos \lambda_{sl} - (-dF_{Ax} \cos \kappa_{rl} +$

$dF_{Ay} \sin \kappa_{rl}) \sin \lambda_{sl}) \sin \gamma_{nl}$ [2]

where dF_{Ax}, dF_{Ay} and dF_{Az} are the elemental partitioned force components (i.e., after accounting for the edge force components due to a rounded cutting edge) in the X, Y and Z directions respectively. Thus, the resultant frictional force on the rakeface is:

$d\vec{R}' = d\vec{F}_x' + d\vec{F}_y'$ [3]

It can be seen (Figure 8) that in the non-linear nose radius region, the friction force components vary widely.

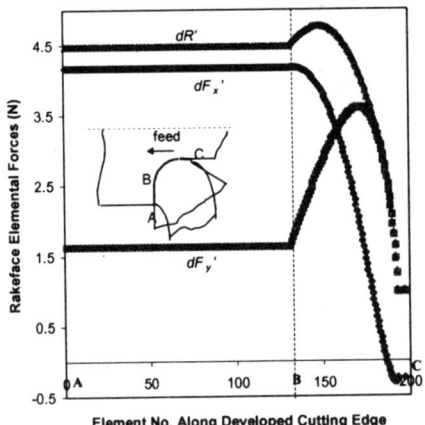

Figure 8. *Variation of elemental rakeface force components and resultant force along the developed length of cutting edge.*

The tool-chip interfacial behaviour was studied using the metallographic analysis of chip cross-sections in the transverse direction (i.e. across the chip width). The region close to the tool-chip interface was photographed continuously at different locations across the chip width at a magnification of 500 X to measure the plastic zone thickness values ($\delta t_2'$ and $\delta t_2''$) and grain elongation angle (Γ). Corresponding to the chip cross-section in Figure 6(a), Figure 9(a) shows the variation of the plastic zone thickness values across the chip width and Figure 9(b) shows the variation of the grain elongation angle across the chip width.

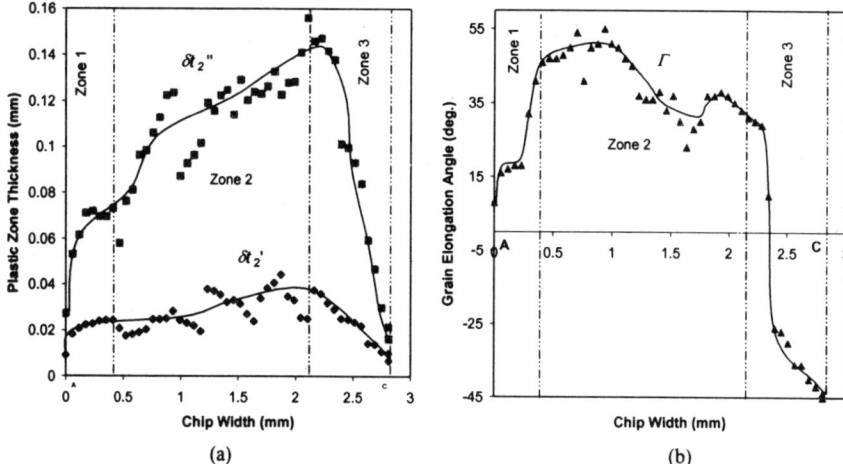

(a)   (b)

Figure 9. *Variation of (a) plastic zone thicknesses (t) and (b) grain elongation angle across the chip width* (Nominal tool geometry: inclination angle = -5 deg., rake angle = -5 deg., side-cutting edge angle = 90 deg.; a = 2.5 mm, f = 0.3 mm/rev., r_ε = 0.8 mm, V = 200 m/min., Work material: AISI 1045 Steel).

The chip transverse cross-section can be demarcated into three zones: (i) Zone 1 is close to the unrestrained end of the depth of cut (A) and exists due to the chip side-spread in this direction; (ii) Zone 2 represents the bulk region of the chip which is influenced by the straight portion of the cutting edge (AB); and (iii) Zone 3 represents the region formed due to the non-linear variations in the nose region (BC). It can be seen that in Zones 1 and 3 the plastic zone thickness values and the grain elongation follow a similar trend as the chip thickness. However, in Zone 2, the chip material is in a state of flux due to the interfering interactions from the chip material flowing from the straight portion of the cutting edge (AB) and the nose region (BC). This increases the frictional conditions in this region causing an increase in the plastic zone thickness and a corresponding decrease in the grain elongation angle. It was seen that the grains are always swept back in the bulk of the cross-section towards the end A (i.e. the free- end of the undeformed area of cut which promotes side- spread of the chip). This is due to two reasons: (i) the end C is a constrained end due to the restriction from the machined surface of the bar which results in material being swept away from C towards A; and (ii) the highly varying elemental chip side-flow in the nose region results in severe interference within the chip material causing a natural flow of the chip material towards the unconstrained end of the depth of cut (i.e. A). However, it was seen that at the end C, the grains tended to flow in a negative direction towards the point C rather than A as in the rest of the chip. This can be understood from the fact that at the end-cutting edge region of the depth of cut, the lead angle is negative thus leading to an opposing flow of chip material. However, this is a highly marginal effect. This analysis was repeated for $a = 1.2$ mm and a similar trend was seen. It was interesting to note that the depth of cut (a) has a role to play on the retarded layer thickness; the smaller depth of cut results in smaller retarded zone thickness. However, the grain orientation angle did not show any significant correlation with the depth of cut.

5.2. Grooved tools

The experiments were repeated for grooved tools at depths of cut of 1.2 and 2.5 mm. Due to the effect of the groove geometry there was reduced chip side-flow and a greater amount of chip back-flow, resulting in up-curl dominated chips which broke after contacting the tool flank. A simple parameter to estimate the contribution of chip up-curl and side-curl in a chip is the twist angle θ (see Figure 10) [NAK 78], [GHO 96]. A purely side-curled chip will have $\theta = 90°$ and a purely up-curled chip will have $\theta = 0°$. The up-curl dominated chip which was produced in the experimental work had an average twist angle of $21°$. Currently ongoing attention focuses on the frictional conditions for these grooved tools. Figure 11 shows the complex tool-chip contact patterns for grooved tools. Metallographic analysis of the transverse chip cross-sections does not show any distinct sweeping back of the grains (i.e. $\Gamma \approx 0$). The longitudinal cross-section on the other hand shows a distinct sweeping back of the grains. Figure 12 shows the cyclic cutting force patterns due to the 3-D cyclic chip formation of these up-curl dominated chips.

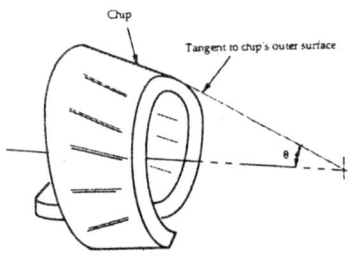

Figure 10. *Definition of twist angle in a 3-D curled mixed mode chip*

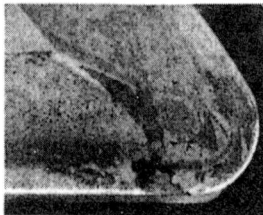

Figure 11. *Tool-chip contact patte* *in 3-D machining with grooved too* (Nominal tool geometry: inclination angle = deg., rake angle = -5 deg., side-cutting ec angle = 90 deg.; $f = 0.3$ mm/rev., $a = 1.2$ m $r_\varepsilon = 0.8$ mm, $V = 200$ m/min., Work mater AISI 1045 Steel, Standard grooved to TNMG 332).

6. Summary

This paper highlights new findings on the role of variable tool-chip interfacial friction in machining. The major points can be summarised as follows:

– The assumption of average tool-chip interfacial friction needs to be critically reviewed in light of the cyclic chip formation processes in 2-D and 3-D machining with both flat-faced and grooved tools.

– In turning, the non-linearity due to the nose radius and the tool geometry imposes variable friction on the tool rakeface along the developed length of the cutting edge giving rise to side-curl. The traditional 2-D orthogonal and oblique theories can predict forces and chip side-flow, but do not account for side-curl.

– Variable friction in 3-D turning operations with flat-faced tools producing predominantly side-curled chips gives a distinct flow of chip material in one direction across the chip width with variability in the plastic zone thickness close to the tool-chip interface and the grain elongation in the chip.

– The degree of up-curl or side-curl domination dictates whether chip material is swept back and retarded across the transverse chip cross-section or not. For more equitable distributions of side-curl and up-curl, a tri-axial plastic flow analysis is necessary in the future work.

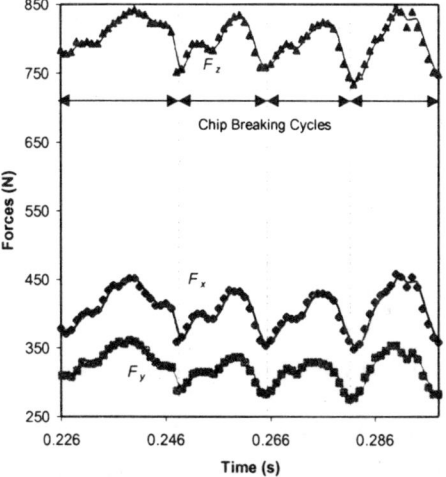

Figure 12. *Cyclic cutting forces corresponding to 3-D cyclic chip formation* (Nominal tool geometry: inclination angle = -5 deg., rake angle = -5 deg., side-cutting edge angle = 90 deg.; f = 0.25 mm/rev., a = 1.2 mm r_ε = 0.8 mm, V = 200 m/min., Work material: AISI 1045 Steel, Standard grooved tool: TNMG 332).

Acknowledgements

The authors acknowledge the support of the National Science Foundation (DMII 9713932) and CRMS (University of Kentucky) for supporting this research work.

110   Friction and Flow Stress in Forming and Cutting

7. References

[BAL 99a] BALAJI A.K., SREERAM G., JAWAHIR, I.S., LENZ. E., "The Effects of Cutting Tool Thermal Conductivity on Tool-Chip Contact Length and Cyclic Chip Formation in Machining with Grooved Tools," *Annals of the CIRP*, Vol. 48/1, p. 33-38, 1999.

[BAL 99b] BALAJI A.K., JAWAHIR, I.S., "Effects of Cutting Parameters on Chip Side-Curl Mechanisms and Variable Tool-Chip Contact in Turning," *Proc. ASME IMECE, Symp. Machining Sc. & Tech.*, MED-Vol. 10, p. 311-318, November 1999.

[BRO 87] BROWN C.A., "Strain Analysis in Machining Using Metallographic Methods," *Metallography*, Vol. 20, p. 465-483, 1987.

[CHA 84] CHALLEN J.M., OXLEY P.L.B., "Some New Thoughts on the Mechanisms of Sliding and Sticking Friction in Metal Working Processes," *Adv. Tech. Of Plasticity*, Vol. 1, p. 127-132, 1984.

[DEW 78] DEWHURST P., "On the Non-uniqueness of the Machining Process," *Proc. R. Soc. Lond.*, Vol. A360, pp. 587-610, 1978.

[DOY 79] DOYLE E.D., HORNE J.G., TABOR, D., "Frictional Interactions between Chip and Rakeface in Continuous Chip Formation," *Proc. R. Soc. Lond.*, Vol. A366, p. 173-187, 1979.

[FAN 96] FANG X.D., JAWAHIR I.S., "An Analytical Model for Cyclic Chip Formation in 2-D Machining with Chip Breaking," *Annals of the CIRP*, Vol. 45/1, p. 53-58, 1996.

[FAN 00] FANG N., JAWAHIR I. S., OXLEY P. L. B. , "A Universal Slip-line Field Model With Non-unique Solutions for Machining With Curled Chip Formation and a Restricted Contact Tool," (Paper accepted for publication) *Int. Journal of Mechanical Sciences*, 2000.

[FIN 56] FINNIE I., SHAW M.C., "Friction Process in Metal Cutting," *Trans. ASME*, Vol. 78, p. 1649-1665, 1956.

[GAN 98] GANAPATHY B.K., JAWAHIR, I.S., "Modeling the Chip-Work Contact Force for Chip Breaking in Orthogonal Machining with a Flat-faced Tool," *ASME Journal of Manufacturing Sc. & Engg.*, Vol. 120(1), p. 49-56, 1998.

[GHO 96] GHOSH R., REDETZKY, M., BALAJI, A. K., JAWAHIR, I. S., "The Equivalent Toolface (ET) Approach for Modeling Chip Curl in Machining with Grooved Tools," *Proc. CSME Forum*, p. 702-711, 1996.

[JAW 95] JAWAHIR I.S., ZHANG, J.P., "An Analysis of Chip Curl Development, Chip Deformation and Chip Breaking in Orthogonal Machining," *Trans. NAMRI/SME*, Vol. 23, p. 109-114, 1995.

[OXL 89] OXLEY P. L. B., *Mechanics of Machining: An Analytical Approach to Assessing Machinability*, Ellis Horwood Ltd., Chichester, United Kingdom, 1989.

[MAD 97] MADHAVAN V., CHANDRASEKHAR, S., Farris, T.N., "Direct Observations of the Chip-Tool Interface in Machining," *Proc. ASME IMECE*, MED 6-2, pp. 45-52, 1997.

[MAT 81] MATHEW P., OXLEY, P.L.B., "Allowing for the Influence of Strain Hardening in Determining the Frictional Conditions at the Tool-Chip Interface in Machining," *Wear*, Vol. 69, pp. 219-234, 1981.

[MER 44] MERCHANT M.E., "Basic Mechanics of the Metal Cutting Process," *Transactions of the ASME*, Vol. 66, pp. A168-A175, 1944.

[NAK 78] NAKAYAMA K., "Basic Rules on the Form of Chip in Metal Cutting," *Annals of the CIRP*, Vol. 27/1, pp. 17-21, 1978.

[SPA 71] SPAANS C., The Fundamentals of Three-dimensional Chip Curl, Chip Breaking and Chip Control, Ph.D. Dissertation, TU Delft, Netherlands, 1971.

[TRE 88] TRENT E.M., "Metal Cutting and the Tribology of Seizure: Part II - Movement of Work Material Over the Tool in Metal Cutting," *Wear*, Vol. 128, pp. 47-64, 1988.

[RED 99] REDETZKY M., BALAJI, A.K., JAWAHIR, I.S., "Predictive Modeling of Cutting Forces and Chip Flow in Machining with Nose Radius Tools," *Proc. 2nd CIRP Int. Workshop on Modeling of Machining Operations*, Nantes, France, pp. 160-180, January 1999.

[WAL 64] WALLACE P.W., BOOTHROYD, G., "Tool Forces and Tool-Chip Friction in Orthogonal Machining," *J. Mech. Engg. Sci.*, Vol. 6(1), pp. 74-87, 1964.

[ZOR 63] ZOREV N.N., "Interrelation Between Shear Processes Occurring Along Tool Face and on Shear Plane in Metal Cutting," *Int. Res. in Prodn. Engg., ASME*, Pittsburgh, PA, USA, p. 42, 1963.

Chapter 8

Sensing Friction: Methods and Devices

J. Jeswiet and P. Wild

Dept of Mechanical Engineering, Queen's University, Ontario, Canada

1. Introduction

Stress at the work-piece/die boundary, in both bulk forming and sheet forming, is arguably the single most important physical parameter influencing the processing of metals, yet it remains the least understood parameter; hence the need for basic research into metal-die interface mechanisms. To gain a good understanding of the mechanisms at the interface and to be able to verify the friction and tribology models that exist, friction sensors are needed.

2. Friction sensors in general

Designing sensors to measure friction in metal working has been pursued by many researchers; see Schey [SCH 83]]. Two types of sensor which are used in laboratories are: (1) the pin-type sensor originally devised by van Rooyen and Backofen [ROO 57] and improved by Al-Salehi [ALS 73] for use in metal rolling and (2) the cantilever friction sensor [BRI 86], see Figure 1. The pin-type sensor {1} measures two forces at the interface, one orthogonal to the rolling direction and the other at an angle to rolling and is specially designed for measuring friction in rolling.; by converting the angular measurement, and comparing it to the normal force, the friction is eventually obtained. The cantilever sensor can be used in other situations, including rolling; it will be discussed in detail later in this chapter.

Other devices which are not reviewed by Schey [SCH 83] but have been tried, are those by Jeswiet [JES 81], see design {3} in Figure 2, Stelson [STE 84], see design {4} in Figure 2, and Li et al [LI 92], see design {5} in Figure 2. The device designed by Li et al [LI 92] is of particular interest because it uses piezo-electric elements which were also of interest to our research program but which have proven to be a disappointment because of temperature sensitivity. The method devised by Stelson [STE 84] never gave any meaningful results and hence was discarded; in effect calibration was impossible as was verification of the results. Design {3}, the device considered by Jeswiet [JES 81], could not be calibrated as shown by photo-elastic analysis at that time, and hence was discarded; however a new sensor configuration similar to that design, called the *diaphragm sensor* is now under development; it is discussed in the following section. The following discussion will centre on the diaphragm friction sensor and cantilever friction sensor.

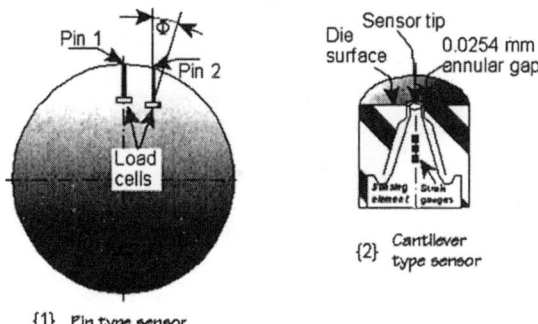

{1} Pin type sensor

Figure 1. *The pin sensor and cantilever sensor*

3. Development of the diaphragm design

Most applications for measuring friction require that the sensor does not disturb the die surface; lubrication is a good example. The need for such a sensor led to a design called the diaphragm sensor and which is being developed. The first embodiment of the diaphragm sensor to be analysed included a central post, as shown in Figure 3, to ensure good load transfer between the work piece and the sensor. Based on finite element modelling, this design was first shown to be potentially feasible and then optimised. Other variations on this design are now being considered.

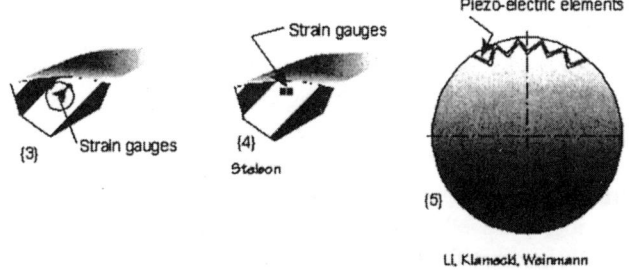

Figure 2. *Other sensors which have been tried, unsuccessfully*

The diaphragm sensor was the result of a brainstorming session, which led to an FE analysis. The finite element analysis, shown in Figure 4, showed the strain distribution on the underside of the diaphragm is asymmetric. Also, the finite element analysis shows the average strains on the left and right of the diaphragm centre differ by approximately 50 microstrain. Conventional strain gauges can resolve 1 microstrain, hence this was an encouraging result, which has led to further proof of concept work.

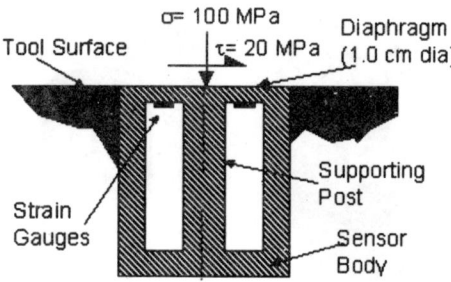

Figure 3. *Schematic of diaphragm sensor*

A first prototype of the diaphragm friction sensor with the supporting post was developed and tested. The diaphragm and post were machined as a single tee-shaped component. Strain gages were applied to the underside of the disc-shaped top of the tee, which was then press fit into a cylindrical recess in the sensor body. The minimum permissible size of the tee was dictated by the size of the smallest commercially available strain gages (2.4 mm x 3.3 mm). This sensor showed good sensitivity to both normal and shear loading, as shown in Figure 4. The data shown is for a single strain gage. By processing the data from two opposed gauges, the magnitude of any normal or shear loading can be resolved.

4. Development of the cantilever sensor

Another sensor under development at Queen's is called a *cantilever sensor.* Development of this sensor has been going on for some time and its genesis is aluminium; up to 25% reduction. However the drawback of this sensor was that it could only measure surface forces for very low reductions and it could only measure two forces, the normal force averaged over the width of the sensor and the friction force in the rolling direction, averaged over the width of the sensor. Design {8} was the first sensor, which could measure forces in all three orthogonal directions [BRI 86]. The gap around the sensor was 0.05 mm, hence there was very little extrusion of aluminium into the gap, and the sensor could be used at higher reductions; however extrusion into the gap did still occur and because of this the sensor could not be used for lubrication studies. Design {9} was an effort to make fabrication of the sensor easier, which it did, and to reduce the gap even further to 0.025 mm; extrusion into the gap still occurred at reductions higher than 20%. In design {10} the idea was to insert indium into the 0.025 mm gap thereby stopping extrusion into the gap. The indium would also be "fluid" enough to allow the sensing element to move back and forth. Tests showed that indium did indeed stop extrusion into the gap, up to reductions of 30%. However at higher reductions extrusion still occurred. The final iteration, design {11}' includes using a steel "cap" on the top of the cantilever sensor [ALI]. With this configuration the effective sensor area is larger than the tip of the sensing element but with proper calibration an "effective" area can be found. This design has the distinct advantage of eliminating extrusion into the gap and allowing the sensor to be used for lubrication studies, which we are now doing. Figure 6 shows the sensor disassembled.

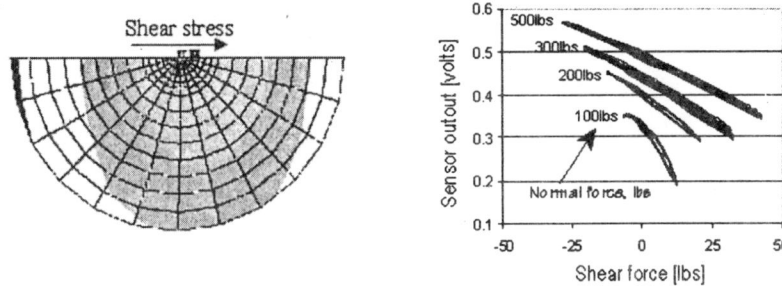

Figure 4. *Preliminary results of diaphragm sensor*

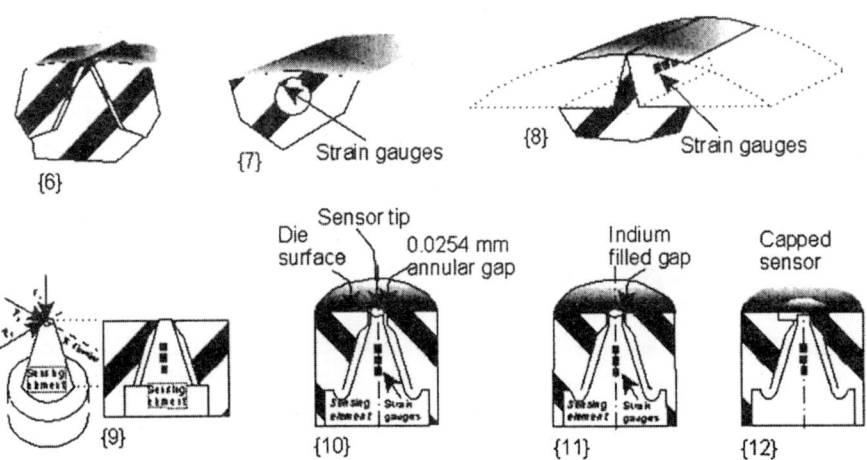

Figure 5. *Genesis of cantilever sensor*

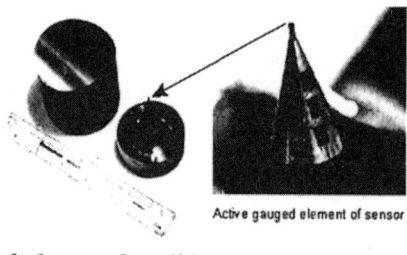

Active gauged element of sensor

Cantilever sensor disassembled

Figure 6. *Photograph of disassembled sensor*

5. Some results

Figures 7 and 8 show some results which were obtained both with the latest "capped-sensor" and without the cap. Results of two sets of experiments are shown for the ring compression test.

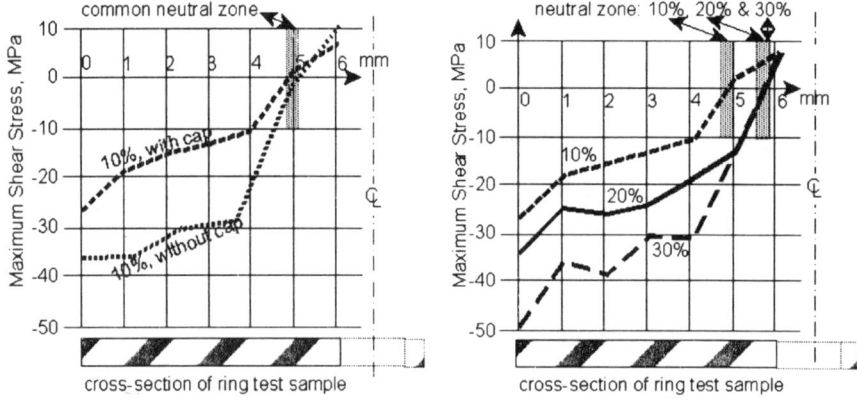

Figure 7. *Ring test measurement for 6061*   **Figure 8.** *Ring test result with capped aluminium; 10% reduction for capped and sensor for 6061 uncapped sensors*

Ring compression tests were conducted with uncapped and capped sensors. In both cases the neutral zone was observed at the same position, as expected for no lubrication; see Figure 7. Hence, with the cap in place we are now in a position to be able to measure the effect of lubricants for different forming processes.

Once confidence had been gained that the capped sensor gave dependable results it was tested with 6061 aluminium ring test samples without any lubricant at reductions of 10%, 20% and 30%. The results of those tests can be observed in Figure 8.

In another set of experiments compression tests were conducted on solid aluminium billets. Coulomb's model for friction was used and Figure 9 shows how friction varies with compression. Results for both the capped and uncapped sensor are shown. These results can be compared with earlier results for an uncapped sensor in cold rolling of aluminium shown in Figure 10. Using Coulomb's model, the results were compared to friction results calculated with the method to calculate friction developed by Ford [FOR 63].

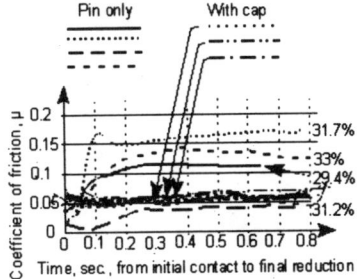

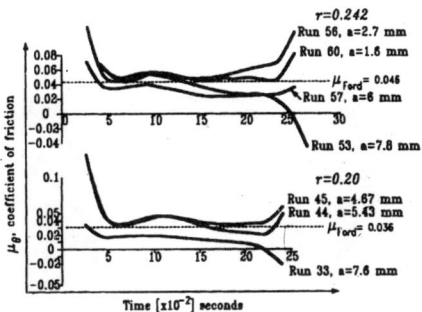

Figure 9. *Variation of coulomb friction in compression of solid 6061 billets; both sensors*

Figure 10. *Variation of Coulomb Friction for cold rolling of 6061 aluminium billets with values calculated by Ford's method*

6. Calibration

An important part of sensor development is the calibration of the device. A special calibration rig has been designed for this purpose. It can apply known combinations of friction and normal loads. Figure 11 is a picture of the calibration rig which has been used for all the sensors developed at Queen's and described in this paper.

Another calibration strategy is to measure friction while the sensor is mounted in a "capstan" rig which is used to test lubricants; this test was developed by Wilson and Malkani [MAL 91]. At present, a variation of the uncapped sensor, design {9}, is being used to measure friction in a capstan testing device at ALCAN Research. Future work will include using a capped cantilever sensor. Using the capped cantilever sensor can be an important feedback tool in experimental rolling mills, which is used to determine lubrication factors for processing mills.

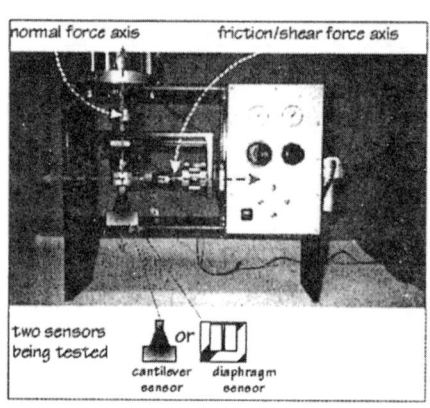

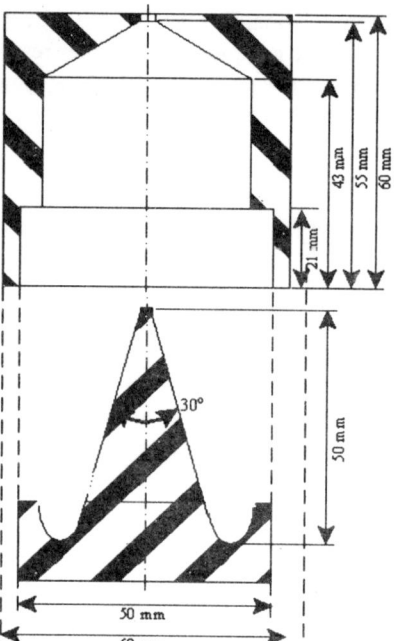

Figure 11. *Friction sensor calibration unit*

Figure 12. *Details of cantilever sensor*

7. Conclusions

A survey of methods used to sense friction in metal forming has been conducted. Of the sensors surveyed, details of the sensors developed at Queen's have been discussed.

The latest cantilever sensor design, the capped cantilever sensor, type {12} shown in Figure 5, gives accurate results and will provide industry and research laboratories with a valuable tool, which can be used to determine lubrication characteristics at contact surfaces and in process monitoring, control. Details of the sensor are given in Figure 12.

The diaphragm sensor, shown in Figure 4, shows promise as a friction measurement device for metal forming. However, it is still at the proof of concept stage and special, micron sized strain gauges are being developed for this design.

Acknowledgements

The authors wish to thank both the Natural Science and Engineering Research Council of Canada and Materials and Manufacturing Ontario for funding this research.

8. References

[SCH 83] SCHEY J.A., *Tribology in Metalworking,*.8, 1983, ASM; p. 204 - 224.

[ROO 57] VAN ROOYEN G.T. AND BACKOFEN W.A., "Friction in Cold Rolling", *J. Iron & Steel Institute*, 1957, p 235.

[ALS 73] AL-SALEHI F..A.R, FIRBANK T.C. and LANCASTER P.R. ,"Experimental determination of the Roll Pressure Distribution in Cold Rolling", *International Journal of Mechanical Science*, 1973, vol 15, p. 693 - 710.

[BRI 86] BRITTEN D. and JESWIET J., "A Sensor for Measuring Normal Forces with Through and Transverse Friction Forces in the Roll Gap", 1986, *SME/NAMRI Transactions*, p. 355 - 359.

[JES 81] JESWIET J., PhD thesis Queen's Mechanical Engineering, 1981.

[STE 84] STELSON K.A., "A New method to Measure Friction and Normal Pressure and Frictional stresses in the Roll Gap during Cold rolling", 1984, *SME/NAMRI Transactions*, p. 259 - 264.

[LI 92] LI M., KLAMECKI B. E., and WEINMANN K.J., "SN-Gauge and Instrumented Pin: Devices to Measure Shear and Normal Tool Forces in Sheet Metal Forming", *Transactions of NAMRI/SME*, v. XX, 1992; p. 103 - 107.

[BAN 72] A. BANERJI A. AND RICE W.B., "Experimental Determination of Normal Pressure and Friction Stress in the Roll Gap During Cold Rolling", *Annals of CIRP 22/1/1972*, p 53.

[ALI] ALI M., "A Sensor for Measuring Surface Shear Stresses in Metal Forming", MSc Thesis Queens University Mechanical Engineering In Press.

[JES 98] JESWIET J., "Friction Forces in Dry Rolling of Aluminium", Sheet Metal, 1998, *Proceedings of International Conf,* Enschede, Netherlands; vol I, p. 315 - 322.

[FOR 63] FORD H., Mechanics of Materials, 8, Longmans, 1963.

[MAL 91] MALKANI W. R. D., SAHA H., Advanced Boundary Friction Measurements Using a New Sheet metal Forming Simulator. Wilson, P.K., *Proceedings of NAMRC XXII*, 1991, p. 37 - 42.

Chapter 9

The Problem of Constitutive Equations for the Modelling of Chip Formation: Towards Inverse Methods

F. Meslin and J.C. Hamann
Mechanics and Materials Laboratory, Ecole Centrale de Nantes, Nantes, France

1. Introduction

One way to increase productivity is to improve the cutting conditions in machining processes. In order to achieve this, one solution is to use new cutting tools such as coated tools or ceramics. The optimisation of tool geometry and cutting conditions is a possible second way. However, such optimisation requires great knowledge of the mechanics of chip formation in metal cutting operations.

Metal cutting modelling provides understanding and prediction of machining process variables such as stress, strain, cutting force. Early cutting models of chip formation were based on the shear plane model [MER 44] or slip-line theory [LEE 51]. These models usually have unknown constants, which must be obtained for each cutting tool/workpiece couple by experimental means. Moreover, these types of models could not provide any means to study the non-linear behaviour of the workpiece material. These kinds of approach produce descriptive models rather than predictive models.

More recent work has used finite element methods in order to include effects of the cutting process such as friction at the tool-chip interface, work-hardening, strain rate and temperature dependencies of the workpiece material response. The finite element method provides a possibility to solve large non-linear problems.

Three kinds of mechanical formulations can be used. The Lagrangian formulation is a "solid" approach. In this case, the grid of the mesh is attached to the workpiece material. So the chip formulation needs to modify the mesh just in front of the cutting edge. A chip separation criterion [SHI 93, STR 85] can be used in order to perform the chip formation. However, we need to make some assumptions about the localisation of the chip formation and we lose all mechanical information in front of the cutting tool. More recent Lagrangian methods in the modelling of chip formation use a remeshing algorithm in order to separate the chip of the workpiece [CER 96, FOU 99]. This interesting method leads to CPU time consumption, especially to reach the stationary solution.

Eulerian method [LEO 99, STR 90] (a "fluid" approach), in which the grid of the mesh is not attached to the material, is computationally efficient and avoid problems of mesh distortion. However, such a procedure requires iterative modification of the free chip geometry [MAE 96]in order to satisfy the velocity boundary conditions.

An alternative method is to use Arbitrary Lagrangian Eulerian (ALE) formulation [PAN 96, DAN 99]. In this case, the nodal points of the mesh are not attached to the material point. However, the nodal points of the mesh are not fixed in the space, but have the possibility to move in order to avoid catastrophic distortions of the mesh during the chip formation. In fact, two sets of coordinate systems are used in the ALE procedure to describe the motion of the material point and the motion of the grid points. The ALE formulation appears as a mix of Lagrangian and Eulerian formulations. Thus, we take benefit from the efficiency of the Eulerian procedure and avoid the problem of the free chip geometry.

However, whatever the formulation used, numerical simulation of chip formation requires (i) material flow characteristics at high temperature, strain-rate and strain as encountered during cutting process, (ii) a tool-chip contact friction. The material flow characteristics are usually obtained from a Split Hopkinson pressure bar bench equipped with a high energy heating device or impact tests [MAE 83]. Unfortunately, these devices cannot provide, at the same time, more than 5.10^3 s^{-1} strain rate and 0.5 plastic deformation. It means that for the constitutive equation to be evaluated an extrapolation will be necessary.

The friction parameters at the tool-chip contact are hardly identified. Only a few methods are available and, in all cases, experiments are not conducted in similar conditions to those encountered in cutting process [JOY 94, OLS 89, HAM 98].

The main objective of this paper is to present results about the classical identification of the Johnson-Cook constitutive relation. In the first part, through examples, we highlight the respective influence of thermal softening, strain hardening and strain rate sensitivity. Then, in the second part, we discuss the possibility to discriminate two steels grades that have the same overall mechanical characteristics but not the same machinability. We discuss also the difficulties to identify constitutive equations for material such as Titanium alloy, which exhibits specific thermal properties. A last, we propose some ways of investigation in order to improve the identification of constitutive relation dedicated to the modelling of chip formation in metal cutting.

2. Johnson-Cook formulation and rheological parameters

2.1. Constitutive equation

Numerical simulation of chip formation requires a thermo-visco-plastic law. In the case of large plastic deformations and large strain rates, the well known Johnson-Cook [JOH 83] formulation is often used. So, the flow stress σ_{eq} is given by:

$$\sigma_{eq} = (A + B\varepsilon^n) \left[1 + C \ln \left(\frac{\dot{\varepsilon}}{\dot{\varepsilon}_0} \right) \right] \left[1 - \left(\frac{T - T_{amb}}{T_f - T_{amb}} \right)^m \right] \qquad [1]$$

The above expression gives the flow stress (σ_{eq}) as a function of the effective plastic strain (ε), the effective plastic strain rate ($\dot{\varepsilon}$) and the temperature (T). $\dot{\varepsilon}_0$, T_{amb} and T_f are respectively the reference strain rate, the room temperature and the melting temperature of the material. A, B, C, n and m are rheological parameters to be identified.

The three parts of the Johnson-Cook formulation define respectively the strain hardening, the strain rate sensitivity and the thermal softening of the material.

2.2. Evaluation of the constitutive equation

The evaluation of a Johnson-Cook law requires specific means, i.e. a split Hopkinson pressure bar bench (SHPB) equipped with a high energy heating device. This technique is certainly the most often used because of versatility. Basically two types of loading are used: torsional tests or compressive tests.

A first set of tests is carried out in the case of quasi-static conditions (strain rate lower than 10^{-2} s^{-1}). In these conditions of loading, the temperature of the material sample is the same as the room temperature (isothermal conditions). A second set of tests is conducted for large strain rate (strain rate greater than 10^3 s^{-1}). This time, high strain rates lead to adiabatic conditions.

2.3. Relative effects of the rheological parameters

In order to appreciate the relative influence of the strain hardening, thermal softening and strain rate sensitivity that appear in the Johnson-Cook law, we propose to analyse the constitutive equation obtained for a 42CrMo4 (AISI 4142) steel grade. Rheological parameters identified by classical means (split Hopkinson pressure bar bench associated to classical compressive tests) for this steel grade are given in Table 1. More details about the identification could be found in [GRO 96].

Material	A (MPa)	B (MPa)	n	C	m	$\dot{\epsilon}_0$
42CrMo4	598	768	0.2092	0.0137	0.807	0.001

Table 1. Constitutive equation coefficients of 42CrMo4 steel grade identified by classical means [GRO 96].

Figure 1 shows the combined influence of strain hardening, temperature softening and strain rate sensitivity in the case of a 42CrMo4 steel grade. Even if the strain rate sensitivity is obvious, the variation range covers 6 decades, i.e. from 10^{-3} to 10^3 s^{-1}, while the strain hardening is concerned with a 1 decades range (from 0.1 to 1) and the temperature by half a decade variation (from 293 K to 1293 K). It follows that the thermal softening is the most important phenomenon in terms of potential stress variation, followed by the strain hardening and finally by the strain rate sensitivity.

However, the strain rate influences to a large extent the thermal conditions of the test. In quasi static conditions, i.e. for a 10^{-3} s^{-1} strain rate, the test is carried out under isothermal conditions and the specimen is kept at room temperature during the test. For a 10^3 s^{-1} rate, the test is performed under adiabatic conditions and the sample temperature increases, thus "softening" the material.

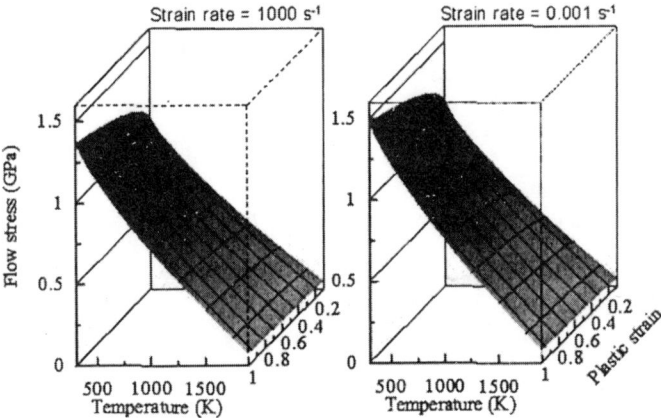

Figure 1. *Flow stress evolution of a 42 CrMo4 steel grade according to the strain rate, temperature and plastic deformation.*

3. Difficulties in the identification

3.1. *Case of a Titanium alloy*

Titanium alloys are known for their great sensitivity to shear localization [MAC 90, KOE 79]. Depending on the experimental device used to identify the constitutive relation, one can obtain a large variation in rheological parameters. For example, a Titanium alloy has been identified by two devices: (i) a classical bench, (ii) a bench equipped with a high energy heating device [LES 96]. In the case of the identification made by a classical device, the Johnson-cooling law is used. Identified rheological parameters are given in Table 2.

Material	A (MPa)	B (MPa)	n	C	m	$\dot{\epsilon}_0$
Titanium alloy	0.9	1	0.03	0.275	4	0.0935

Table 2. *Johnson-Cook coefficients for Titanium alloy identified by classical means.*

In the case of the bench equipped with a high energy heating device, a modified Johnson-Cook law gives the best results to approximate the behaviour of a Titanium alloy. The flow stress (σ_{eq}) is written as follows:

$$\sigma_{eq} = \sigma_0(1 + \varepsilon^n) \left[1 + \left(\frac{\dot{\varepsilon}}{\dot{\varepsilon}_0} \right)^{1/p} \right] \left[\frac{e^{-T/T_0}}{e^{-T_{amb}/T_0}} \right] \qquad [2]$$

In the above expression, ε is the effective plastic strain, $\dot{\varepsilon}$ the effective plastic strain rate, T the temperature and T_{amb} the room temperature. The other parameters (T_0, $\dot{\varepsilon}_0$, σ_0, n, $1/p$) must be identified by experimental tests. Table 3 gives the rheological coefficients identified with a bench equipped with a high energy heating device.

Material	σ_0(MPa)	$\dot{\varepsilon}_0$ s^{-1}	n	$1/p$	T_0 (K)
Titanium alloy	1025	$57{,}5.10^3$	0.37	0.424	1170

Table 3. *Rheological coefficients for Titanium alloy identified with a bench equipped with a high energy heating device [LES 96].*

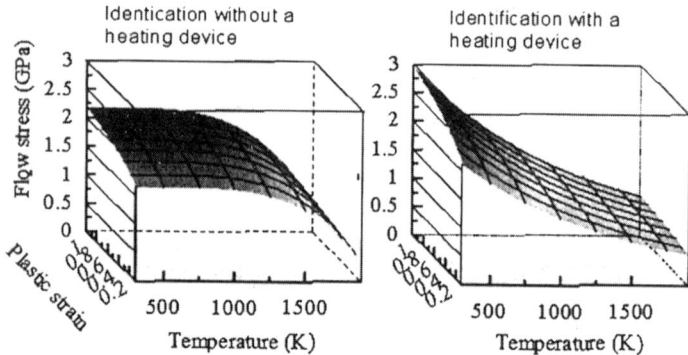

Figure 2. *Comparison of the constitutive equation of a Titanium alloy evaluated on a split Hopkinson bar bench equipped or not with a heating device. The strain rate has been fixed at 10000 s^{-1}.*

Figure 2 shows the evolution of flow stress according to the temperature and the strain, for the two types of identification. We can notice that the behaviour obtained by the classical bench is very similar to that of the 42CrMo4 steel grade. However, we obtain a specific evolution of flow stress in the case of a material identified with a bench equipped with a high energy heating device.

Actually the problem with Titanium alloys is that their sensitivity to shear localisation is so high that the phenomenon occurs during the compressive test. As a consequence, the plastic strain reached during the test is very low and the strain energy dissipated too low to heat up the specimen enough and to reveal the temperature softening of the alloy.

3.2. *Influence of the inclusion content*

Machinability improvement treatment allows an increase in cutting conditions such as cutting speed or the tool life. For low loading, treated steel grade exhibits generally similar mechanical properties to the "classical" steel grade.

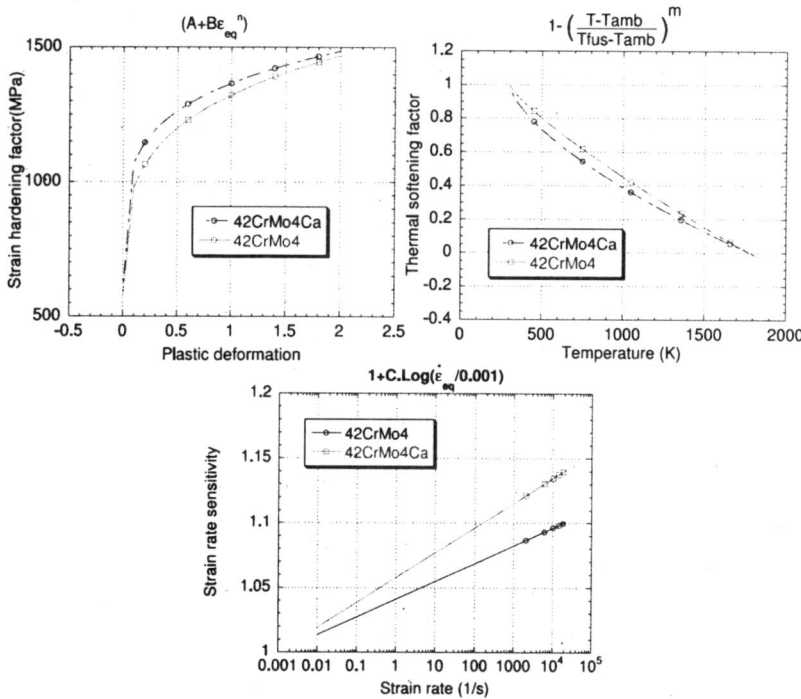

Figure 3. *Comparison of the behaviour of a standard and a calcium treated 42CrMo4 steel.*

In order to know if a classical identification of constitutive relation allows one to discriminate two similar steels (treated and untreated), we propose to analyse the case of two grades of 42CrMo4 steels named: 42CrMo4 and 42CrMo4Ca. These two steel grades have the same metallurgic base but the 42CrMo4Ca has been treated by calcium in order to improve its machinability. Thus, we obtain two steels that exhibit similar mechanical properties but not the same machinability. For these two steels, the rheological parameters of the Johnson-Cook law are given in Table 4. More details about the identification can be found in [GRO 96].

Figure 3 shows respectively the three factors of the Johnson-Cook law (strain hardening, thermal softening and strain rate sensitivity) for the two steels. We notice

Materials	A (MPa)	B (MPa)	n	C	m	$\dot{\epsilon}_0$
42CrMo4	598	768	0.2092	0.0137	0.807	0.001
42CrMo4Ca	560	762	0.2555	0.0192	0.660	0.001

Table 4. *Constitutive equation coefficients of 42CrMo4 and 42CrMo4Ca steels identified by classical means [GRO 96, HAM 98].*

that the behaviours of the 42CrMo4 and the 42CrMo4Ca are very similar. The effect of the calcium treatment is hardly identified by the classical split Hopkinson bar bench

The greater difference in terms of stress level is observed in the medium temperature range, i.e. between 500 and 700 K, for high strain rate and for strain. However, even if this loading is similar to that observed in chip formation, could this difference be considered as significant?

3.3. *Case of austenitic stainless steels*

Stainless steels of the 304 family undergo a strain induced martensitic phase transition when subjected to high strain rates. The transition plasticity is mixed with classical plastic deformation, but the response of each deformation mechanism changes with the strain rate. The effect of this transition plasticity is the appearance of stress-strain curves with two strain hardening portions, whose relative importance depends on the strain rate but also on the rolling ratio of the workpiece from which the samples are extracted. As explained in Figure 4, the Johnson-Cook formulation cannot represent a two strain hardening portions curve and a compromise has to be found.

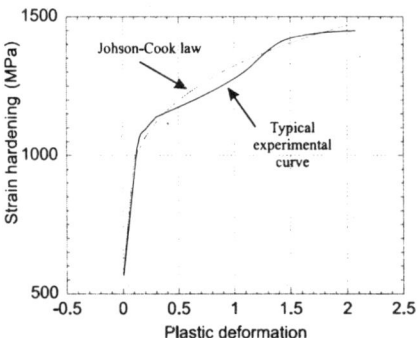

Figure 4. *Typical strain hardening portions obtained by experimental way. The Johnson-Cook law can not represent this specific experimental curve*

Usually, since the machining process involves large plastic strains, the second part corresponding to the predominance of plasticity will be given much attention. However, since the tests conducted on a Hopkinson bar bench allows a maximum deformation of about 0.5 (at high strain rates) the part of the first portion cannot be neglected, because of the heat generation associated with this portion, which is lower compared to a standard plastic deformation process.

Thus, according to the part of the curve considered to establish the identification, we can obtain two sets of potentially equivalent coefficient candidates, in terms of goodness of fit (Table 5).

Materials	A (MPa)	B (MPa)	C	n	m	$\dot{\epsilon}_0$
Set 1	0.285	2.69	0.047	0.8	0.21	0.001
Set 2	0.285	2.69	0.039	0.591	0.34	0.001

Table 5. *Two sets of potentially equivalent coefficient candidates in terms of goodness of fit*

However, the two Johnson-Cook laws give different results when an extrapolation is made in the high strain rate area. Consequently, we are confronted with the problem of extrapolation and its real impact on flow stress (see section 3.4).

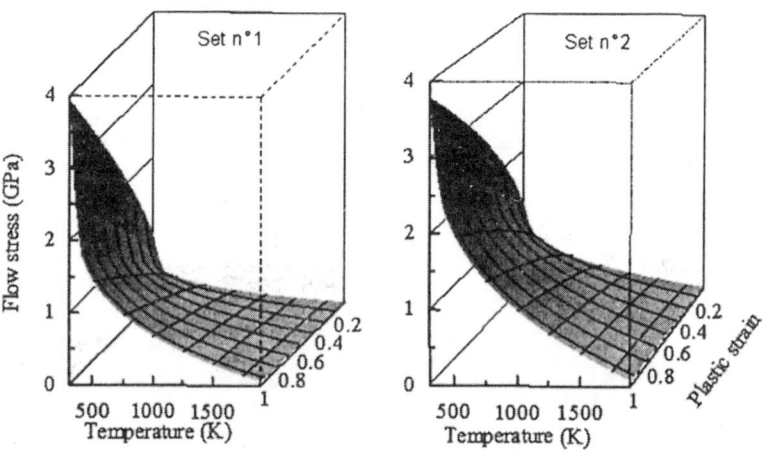

Figure 5. *Evolution of the flow stress for the 2 sets of Johnson-Cook coefficients. The strain rate has been fixed to $10000 \ s^{-1}$*

Figure 5 compares the two constitutive equations according to the thermal softening and strain hardening contributions; as it can be seen these differences are very low in the tested range. In addition, these differences have a similar effect on flow stress that effect the inclusion contents. Thus, if we are able to discriminate between two steels that have the same mechanical properties but not the same machinability, then we make an error in the identification of stainless steels. In other case, if the error is not significant during the identification of stainless steel flow stress, then we are not able to discriminate the effect of the inclusion contents (improvement of machinability).

3.4. The effect of extrapolation

Generally, the bench cannot provide more than 5.10^3 s^{-1} as a strain rate and 0.3 to 0.5 as plastic deformation on "massive" specimens. However, in the case of machining, the strain rate in the primary shearing area reaches values of up to 10^4 - 10^5 s^{-1}, with a plastic deformation higher than 1 [OXL 89].

So, when we use a Johnson-Cook law identified by classical means for the modelling of chip formation, a part of the law is implicitly extrapolated. At this stage, the problem is to know what is the real impact of the extrapolation.

To illustrate this comment, the classical mathematical function $y = \sin x$ has been plotted. According to Figure 6, we can describe the **sine** function either by the **x** function ($y = x$) or by a more complex equation such as $y = -0.043 + 1.3x - 0.41x^2$. Result of extrapolation is obvious. The **x** function appears to be a good approximation of the **sine** function only for low value of $x (x<1)$. In the case of the more complex equation $y = -0.043 + 1.3x - 0.41x^2$, the approximation of the **sine** function is also good for large value of x. So if we use the **x** function as a modelling of the **sine** function (identified for low values of x), we can appreciate the error generated for large values of x. This case is unusual, and we have the similar ambiguity in the case of

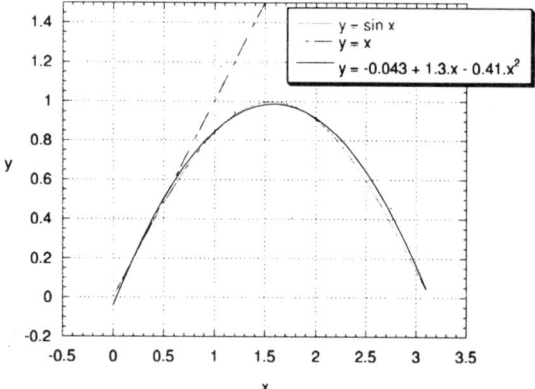

Figure 6. *Example of extrapolation of the sine function*

the use of Johnson-Cook formulation identified by classical means, for the modelling of chip formation.

4. Discussion

From the preceding sections it appears that a constitutive equation evaluated by means of classical compressive tests on a split Hopkinson pressure bar device can be suitable for the modelling of machining but only as a starting point. The major drawbacks are the following:

- The strain rate and the reachable plastic strain are too low
- The thermal softening is not correctly taken into account

In order to overcome these difficulties several ways are possible, but the range of solutions is limited by the principle of the SHPB testing and by the physics of the phenomena. Basically it is a scaling problem but the scaling notion extents over space and time.

4.1. Increasing the strain rate

It is possible to reach high strain rates with the use of loading systems other than compression. Generally speaking it appears that all of these solutions (double shearing, impact plates ...) lead either to low plastic strains or to a very small affected volume. Paradoxically, the fact that only a small volume (a few material grains) is affected by the strain is not an advantage in the scope of chip formation modelling. Actually, the dimension of the area concerned by the primary shearing is very small and can be compared with the grain size of the material. But the tool travels through a long distance in the workpiece and the cutting edge crosses a large volume of work-material. The effects that we want to model are, in a great majority of cases, mean effects, on the mesoscopic scale, and this is one of the major problems with constitutive equation evaluation. The work-material volume that is affected by the instantaneous primary shearing is very low, i.e., a few material grains. The fuzzy layer shared by the chip and the tool that is subjected to secondary shearing, wear and chemical interaction is even smaller in thickness. The average instantaneous volume that will be transformed into a chip can be compared to an elemental chip segment and corresponds to ten up to hundred of grains depending on the material. The size scale of these phenomena is more or less homogeneous. However, they correspond to very different kinetics. The instantaneous mechanical loading is in the μs range of duration, the elemental segment formation is a periodic phenomenon in the range of 10 to 100 kHz while heat flux are a combination of periodic and continuous phenomena in the 10^{-3} to 10^{-1} seconds range and finally chemical interactions are in the 10^{-2} up to several seconds range. Thus, depending on the phenomenon we are interested in, the significant volume that has to be considered in order to ensure a kind of statistical validity is very different.

4.2. *Increasing the specimen temperature*

The case of austenitic stainless steels is a good illustration. The problem of kinetics is once again the major problem. Phenomena such as transition induced plasticity involve short range atomic displacements and can be triggered or inhibited by the temperature whatever the heating rate. On a SHPB bench, and for a 1000 mm length projectile made of steel, the loading pulse duration is 197.10^{-6} seconds. The temperature of the stainless steel specimen reaches 140°C. The average heating rate is therefore very high, about $700.10^{3}°C.s^{-1}$. With so high a heating rate, only a martensitic type of transition can be triggered. Using a high energy heating device it is difficult to control heating rates over $1000°C.s^{-1}$. During machining very high heating rates are caused by instantaneous plastic deformation while lower rates at lower temperatures result from thermal conduction from high to low temperature areas of the work piece. These last heating rates can trigger metallurgical transition involving long range atomic displacements. As a consequence, reproducing the thermal loading is very difficult.

5. Towards inverse methods

5.1. *Framework*

Considering the previous observations it appears that machining is perhaps the only way to recreate the loading conditions found during the chip formation. However, the loading system during chip formation is complex and the response of the work-material is mixed with phenomena such as friction that do not lead to material deformation but that affect, to a large extent, macroscopic variables that can be reliably measured such as cutting force, and also state variables such as temperature.

A solution to obtain a "good" identification of the constitutive relation is to use inverse methods. The aim of these methods is to identify the rheological parameters of the constitutive equation with the help of orthogonal cutting experiments associated with finite element methods simulations. However, this technique requires a lot of experiments and cutting tests. Furthermore, it is difficult to make a distinction between the effects of workpiece material flow, the effects of tool-chip contact interface and the numerical approximations [OZE 98].

On the other hand, the friction parameters at the tool-chip contact are hardly identified. Only a few methods are available and, in all cases, experiment conditions are not conducted in similar conditions as encountered in the cutting process. Pin-on disc friction tests allow only light pressure (< 1Mpa) and temperature. Moreover, the working surface is not refreshed as in the cutting process. So, the value obtained by this means is usually overestimated. The modified pin-on disc device proposed by Olsson [OLS 89] allows one to refresh the working material such as the cutting process. However, the pressures applied on the pin-on disc device are still small regarding the mechanical conditions during machining. The test proposed by Joyot [JOY 94] is made of a

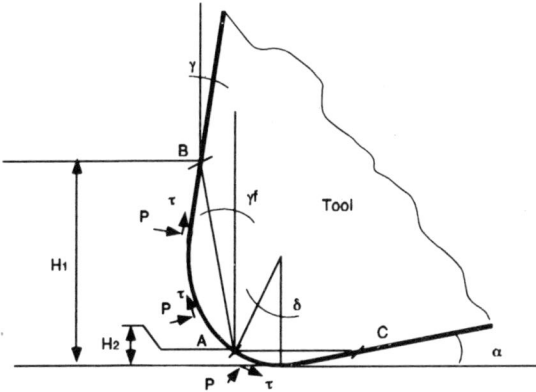

Figure 7. *Definition of the contact area in orthogonal cutting*

frictional tool, a cutting tool and a spring loading system. The spring loading system allows one to impose pressure to the frictional tool, while the cutting tool refreshes the working material in front of the fictional tool. However, even if the loading is high enough, the temperature at the interface between the frictional tool and the working material are not similar with regards to that supposed during the cutting process. So the chemical diffusion at tool-chip interface is not taken into account.

Thus, the major challenge is to discriminate between these phenomena from the material response in order to improve existing inverse methods [OZE 98] and to evaluate the material flow stress response. The next sections propose two approaches to attempt to discriminate the part of friction phenomena during cutting tests. A first approach is based on a restricted cutting tool while the second approach is derived from the experimental observation of the cutting forces.

5.2. Basic experimental principle

The tool-workpiece contact can roughly be separated in two types of phenomena, (i) those that create plastic deformation of the machined material, (ii) those that disperse energy without increasing the plastic deformation of the material, e.g. friction. In order to separate the effects of friction and the effect of plastic deformation, the idea is to change their relative importance. At this point, there are, at least, two possibilities. The first one consists in changing the relative importance of these phenomena on the dimensional scale. For the same cutting conditions, the energy dissipated by the plastic deformation is kept constant while the friction energy in changed. According to Figure 7 this can be obtained by the use of restricted contact length tools. Two types of tool grinding can be used for this purpose as shown in Figure 8. The first case can

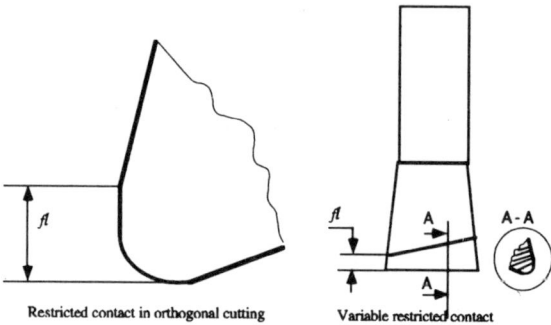

Restricted contact in orthogonal cutting   Variable restricted contact

Figure 8. *Example of tools with restricted contact length for orthogonal cutting or with variable restricted contact for oblique cutting*

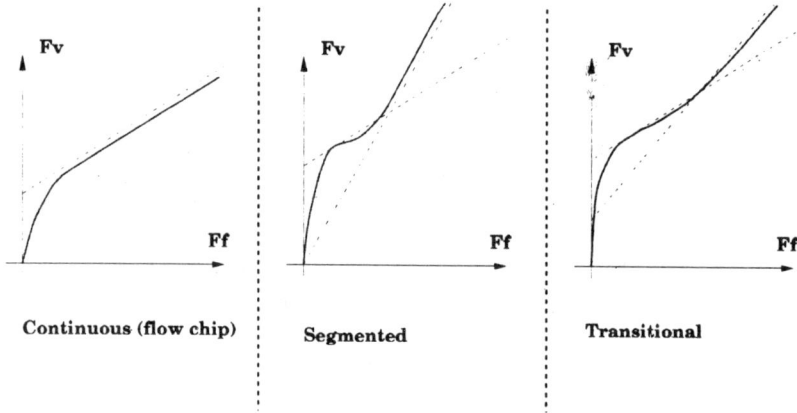

Continuous (flow chip)   Segmented   Transitional

Figure 9. *Typical evolution of $F_v - F_f$ curve according to the chip formation mechanism*

be analysed by an orthogonal cutting model, the second case is more complex because it requires a third dimension (variable restricted contact length).

Another solution is derived from the experimental observation of the cutting forces measured during a cutting test. Basically, the two components of cutting forces (F_c, F_f) during orthogonal turning experiments were measured. For each feed rate, we can plot a point in the (F_c, F_f) plane. As explain in a previous paper [HAM 98], the "$F_v - F_f$" curves change according to the change in the chip formation mechanism. For an equivalent uncut chip thickness the characteristics of the tool-work-material contact are changed when the chip becomes of the "shear localised" type. Comparing results obtained for continuous or segmented chips corresponds to a change in the relative importance of phenomena in the time scale. The two approaches can be mixed.

5.3. *Study of the (F_c, F_f) curves*

When the chip is of the continuous type the relative velocity between the tool and the workpiece close to the edge is 0 (the sticky contact area). The flow velocity gradient spreads across the chip thickness and along the tool chip contact over the rake face. The existence of this gradient gives rise to a dead zone and explains the dissociation and the transfer to the rake face of inclusions that exhibit a strain incompatibility with the workpiece material matrix [HEL 95]. When the chip becomes segmented the phenomenon is periodic. The material removed behaves like rigid blocks, and the chip side of the tool-workpiece contact consists in a thin shear band with a thickness of the same order of magnitude than the primary shear band. For the shear instability to occur, the segment to be formed must be stuck on the rake face during the compression phase until the instability appears thus releasing the segment. The fact that no distortion of the initial micro-structure of the workpiece material is observed in each segment when the chip is fully segmented confirms the lack of flow velocity gradient. In the case of a continuous chip formation, the flow velocity gradient is created because of the simultaneous existence of a sticking contact area and a gliding contact area. In the case of a fully segmented chip, the location of these areas is more or less the same but they alternate with time. With the disappearance of the flow velocity gradient, the behaviour of the resultant cutting force changes according to the uncut chip thickness evolution.

Figure 9 shows the evolution of the Fv-Ff curve with the chip formation mechanism, generally as the cutting speed increases. The linear portion of the curve corresponding to the segmented chip formation crosses the origin when dragged to $Fv = 0$. As a matter of fact when the Fv-Ff curve crosses the origin, the behaviour complies to Merchant's model with a perfectly sharp tool. In case of a change in the chip formation mechanism under specific conditions it is desirable to identify the following variables:

– The critical uncut chip thickness for the transition to occur depending on the cutting speed
– The tool chip friction when segmented chips are formed
– The modification of the contact geometry (or equivalent geometry)

5.4. *Restricted contact length solution*

The deformation zones in orthogonal cutting can be described such as in Figure 10. Three specific zones are usually highlighted: the primary shear zone or the chip formation zone (zone 1), the tool-chip contact or secondary shear zone (zone 2) part of intense friction and lastly, the tertiary deformation zone (zone 3).

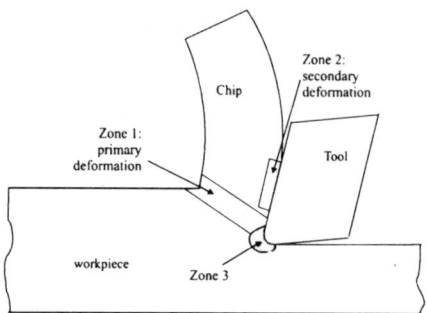

Figure 10. *Deformation zones in orthogonal cutting*

In a first basic approach, one can consider that the dissipated energy during the chip formation in the cutting process is due to a plastic dissipated energy in the primary shear zone (zone 1) and a contact dissipated energy in the secondary shear zone (zone 2). Dissipated energy by zone 3 is not taken into account. So if we are able to modify the contact dissipated energy without any change in the primary shear zone then we are able to highlight the tool-chip contact friction. To obtain this effect, we propose to use a specific designed tool based on restricted contact length (see Figure 8).

The idea is to use a variable contact length of the tool-chip interface along the cutting edge. So, in this case, a part of the chip "glides" without any restriction along the tool rake face, while another part of the chip has a restricted contact at the tool-chip interface. Thus, the first part of the chip (part of chip formed without restriction) "supports" the second part of the chip. In this case, the size of the primary shear zone remains constant along the cutting edge.

The dimension of the rake face changes along the cutting edge. So, with this technique, we can obtain a variable tool-chip contact and keep a large and constant primary shear zone. In other words, we try "to work" the friction part.

Using the variable restricted contact solution implies the use of oblique cutting conditions and simple analytical models can no more be used for the inverse method. A possible overall procedure for this solution is given in Figure 11.

6. Conclusion

The problems of the determination of constitutive relation dedicated to chip formation modelling have been shown. We have highlighted two major key points. First, classical experimental means of identification do not provide similar loading as encountered in the machining process. Secondly, Johnson-Cook's law, which is usually used for the modelling, cannot reflect the specific properties of some materials such as stainless steels.

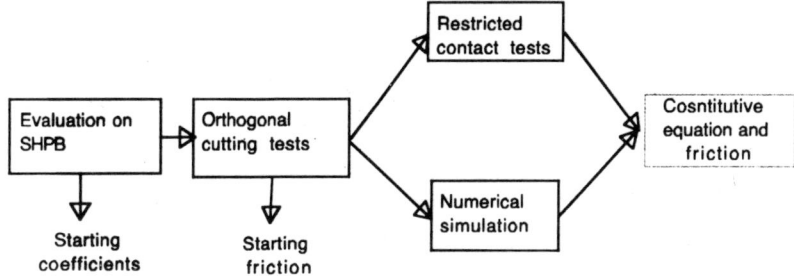

Figure 11. *Overall procedure for the identification of constitutive equation and friction*

For these reasons and according to several authors, we think that only machining gives a reliable approach to the multiscale nature of the load during chip formation. At this stage, a possibility to obtain a realistic identification of the constitutive equations is to use cutting tests associated to numerical simulation: inverse methods.

However, we cannot dissociate two majors effects during a cutting test. A first effect is related to the plastic deformation. Thus, this effect depends on the flow stress response of the workpiece material. The second effect is related to the friction at tool-chip interface. So, when we measure the mascrocopic variables such as chip thickness, force, during a cutting test, we obtained a mix of the two effects described before. Thus, inverse identification of the constitutive relation is not performed under optimal conditions.

In order to improve the determination of the constitutive equation, we propose to discriminate between the part of the flow stress and part of the friction. For this, two methods dedicated to highlight the part of friction at tool-chip interface have been introduced. The first deals with the study of the measured cutting force. The second method proposes to use a restricted contact length tool in order to "do work" the friction.

7. References

[CER 96] CERETTI E., FALLBOHMER P., WU W., ALTAN T., "Application of 2D FEM to chip formation in orthogonal cutting", *Journal of Material Processing Technology*, vol. 59, 1996, p. 169-180.

[DAN 99] DANG E., "Simulation numérique de la formation du copeau", Master's thesis, Ecole Centrale de Nantes, 1999.

[FOU 99] FOURMENT L., "Numerical simulation of chip formation and crack propagation during non steady cutting process", *IInd international CIRP Workshop on Modeling of Machining Operations*, 1999.

[GRO 96] GROLLEAU V., "Approche de la validation expérimentale des simulations numériques de la coupe avec prise en compte des phénomènes locaux à l'arête de l'outil", PhD thesis, Ecole Centrale de Nantes, 1996.

[HAM 98] HAMANN J., LE MAÎTRE F., "Observation and simulation of edge effects at conventional and high speed cutting", *First CIRP International Workshop on Modeling of Machining Operations*, , 1998, p. 299-315.

[HEL 95] HELLE A., "On the intercation between inclusions in steel and the cutting tool during machining", PhD thesis, Acta Poytechnica Sçandinavia, 1995.

[JOH 83] JOHNSON G., COOK W., "A constitutive model and data for metals subjected to large strains, high strain rates and high temperatures", *7th International Symposium on Ballistics*, 1983, p. 541-547.

[JOY 94] JOYOT P., "Modélisation numérique et experimentale de l'enlèvement de matière. Application à la coupe orthogonale", *Doctorat de l'Université de Bordeaux I*, , 1994.

[KOE 79] KOENIG W., "Applied research on the machinability of titanium and its alloys", *Proc. Of 47th meeting of AGARD Struct. And Mat.*, , 1979, p. 1.1-1.10.

[LEE 51] LEE E., SHAFFER B., "The theory of plasticity applied to a problem of machining", *Journal of Applied Mechanics*, vol. 73, 1951, p. 404-413.

[LEO 99] LEOPOLD J., SCHMIDT G., "Challenge and problems with Hybrid Systems for the modelling of machining operations", *II CIRP international Workshop on Modelling of Machining Operations*, , 1999, p. 298-311.

[LES 96] LESOURD B., "Etude et modélisation des mécanismes de formation de bandes de cisaillement intense en coupe des métaux. Application au tournage assisté Laser de l'alliage de titane TA6V", *Doctorat de l'Université de Nantes - Ecole Centrale de Nantes*, , 1996.

[MAC 90] MACHADO A., WALLBANK J., "Machining of titanium and its alloys", *Proc. Of the Inst. Of Mech. Eng. B, Journ. of Eng. Manufacture*, vol. 204, 1990, p. 53-60.

[MAE 83] MAEKAWA K., SHIRAKASHI T., USUI E., "Flow stress of low carbon steel at high temperature and strain rate (Part 2)", *Bull. Japan. Soc. Of Prec. Engr.*, vol. 17/3, 1983, p. 167-172.

[MAE 96] MAEKAWA K., SHIRAKASHI T., "Recent progress of computer aided simulation of chip flow and tool damage in metal cutting", *Journal of Engineering Manufacture*, vol. 210, 1996, p. 233-242.

[MER 44] MERCHANT E., "Basic mechanics of the metal-cutting process", *Journal of Applied Mechanics*, vol. 66, 1944, p. 168-175.

[OLS 89] OLSSON M., "Simulation of cutting tool wear by a modified pion-on disc test", *Int. J. Mach. Tools Manufact.*, vol. 38/1, 1989, p. 113-130.

[OXL 89] OXLEY P., *Mechanics of machining, An analytical Approach to assessing machinability*, Halsted press, 1989.

[OZE 98] OZEL T. ALTAN T., "Modeling of High Speed Machining Processes for Predicting Tool Forces, Stresses and Temperatures Using FEM Simulations", *First CIRP International Workshop on Modelling of Machining Operations*, , 1998, p. 225-234.

[PAN 96] PANTALE O., "Modélisation et Simulation Tridimensionnelles de la Coupe des Métaux", PhD thesis, Université de Bordeaux I, 1996.

[SHI 93] SHIH A., TANG H., "Experimental and finite element predictions of residual stresses due to orthogonal metal cutting", *International Journal for Numerical Methods in Engi-*

neering, vol. 36, 1993, p. 1487-1507.

[STR 85] STRENSKOWSKI J., CAROLL J., "A finite element model of orthogonal metal cutting", *Journal of Engineering for Industry*, vol. 107, 1985, p. 346-354.

[STR 90] STRENSKOWSKI J., MOON K., "Finite element prediction of chip geometry and Tool/Workpiece Temperature Distributions in Orthogonal Metal Cutting", *Journal of Engineering for Industry*, vol. 112, 1990, p. 313-318.

Chapter 10

Rheological Behaviour in Multi-Step Hot Forging Conditions

Paolo F. Bariani, Guido Berti, Stefania Bruschi and Tommaso Dal Negro
DIMEG, University of Padua, Italy

1. Introduction

Material data and constitutive equations together with software processors and validation procedures are the three elements that equally contribute to a successful application of process simulation technology. The accuracy of numerical simulations of hot forming operations strongly depends on the quality in describing the rheological behaviour of the material under deformation.

The material response to hot deformation conditions is affected not only by the current values of the process parameters – strain, strain rate and temperature – but also by the previous thermal and mechanical cycles before the current deformation step, as well as by the variations of temperature and strain rate inside the deformation step.

The constitutive equations currently implemented in commercial FE codes e.g. Norton-Hoff or Arrehnius laws correlate the material flow strength only to the instantaneous values of the process parameters. These equations usually give proper description of the material rheological behaviour only when strain hardening phenomena are dominant. Instead, they do not give reliable results when the thermally activated softening phenomena prevail over the hardening and when the history of deformation cannot be neglected, such as in multi-step hot forging operations.

The influence on flow strength of prior thermo-mechanical histories has been investigated in several hot deformation studies [ALT 97], [BAR 98], [MAR 80], [OH 95], [RAO 96] [YOS 95], [YOS 94]. Most of them refer to stationary processes such as rolling and extrusion and are primarily aimed at developing simulative experiments and formulating constitutive models that are specific for the operation under investigation.

The work reported in this paper has the twofold objective of (i) analysing the material response in single- and multi-step hot deformation conditions through physical simulation experiments, and (ii) assessing the suitability of the multi-step experiments to provide accurate information on material rheology in multi-hit hot deformation studies.

The experiments presented in the paper refer to three different metals typically used in hot forging, a high carbon steel (C70S), an austenitic stainless steel (AISI 304) and a Nickel-based superalloy (Nimonic 80A).

2. Approach

To analyse the material response to both single and multi-step deformation conditions, physical simulation experiments have been carried out on cylindrical samples of the three materials. The approach in testing and analysing the material response to deformation consists of the five following steps (Figure 1) [BAR 99]:

– physical simulation experiments consisting of a *single-step upsetting* of cylindrical specimens at constant values of temperature and strain rate. These experiments correspond to the tests currently used to obtain flow curves and are aimed at evaluating the sensitivity of the material flow stress to temperature and strain rate;

– physical simulation experiments consisting of a *two-step upsetting* of cylindrical specimens with an intermediate soaking time. The former step of deformation and the soaking time represent the thermo-mechanical history of the material when the second step of deformation starts. These experiments are aimed at evaluating and quantifying the influence of the thermo-mechanical history on the flow strength;

– comparison of the flow strength exhibited by the three materials in the second step of deformation of the two-step upsetting experiments with the flow strength measured in the single-step upsetting experiments carried out with the same process parameters;

– comparison of the material microstructures observed just before the starting of the deformation in the experiments with a single-step upsetting with those observed just before the second step of deformation in the experiments with two-steps;

– identification of a microstructural parameter capable of summarising the effects of the previous history on the current value of the flow stress.

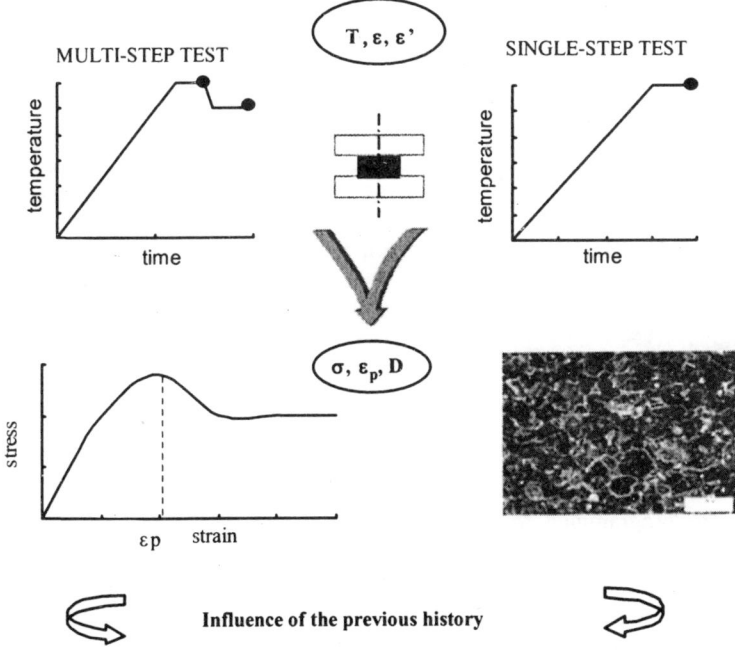

Figure 1. *The approach*

3. Experiments

3.1. Materials

The rheological behaviour of the three following materials has been investigated:
– a high carbon steel, C70S, frequently utilised in forging fracture cap end conrod;

– a stainless steel, AISI 304; applications include chemical equipment, food processing equipment and valves;

– a nickel-based superalloy Nimonic 80A, mostly used in gas turbine engines, steam turbines, and for hot-working tools.

The chemical composition of the three materials is given in Tables 1, 2 and 3, respectively.

Table 1. *Chemical composition of C70S*

Component	C	Mn	Si	Ni	Cr	Mo	P	S	Al
%	0.7	0.49	0.21	0.09	0.15	0.02	0.01	0.06	0.01

Table 2. *Chemical composition of AISI 304*

Component	C	Mn	P	Si	Cr	Ni	S
%	0.08	2	0.045	1	20	10.5	0.03

Table 3. *Chemical composition of Nimonic 80A*

Component	Ni	Cr	Ti	Al	Fe	Co	Mn	Si	C	Cu
%	69	18-21	1.8-2.7	1-1.8	max 3	max 2	max 1	max 1	max 0.1	max 0.2

All the materials have been supplied as 12 mm drawn bars.

3.2. Equipment

All the experiments have been carried out on 12 mm diameter and 14 mm long cylindrical specimens cut off from the bars.

The tests have been run on the thermo-mechanical simulator Gleeble 2000®, which can be programmed to reproduce thermal and mechanical cycles under control of temperature, force, strain and strain rate. The sample is resistance-heated by a thermocouple feedback controlled a.c. current that produces uniform temperature distribution in the diametrical planes (Figure 2 left). The temperature is uniform also along the axis of the specimen thanks to multi-layered interfaces that consist of a sandwich of two alternate foils of graphite and tantalum working, respectively, as lubricant and thermal barrier (Figure 2 right).

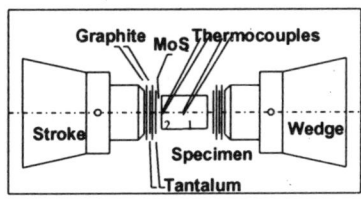

Figure 2. *The working area of Gleeble 2000® system (left) and the multi-layered interface (right)*

3.3. Testing procedures

In the single-step compression experiments (Figure 3a), the specimen is heated at a rate of 15°C/s up to the testing temperature, held at this temperature for 30 s and then compressed at constant strain rate and in isothermal condition to the required amount of strain. During the test, specimen temperature and ram stroke are under computer control, whereas force and strain are continuously monitored. The ranges of variation of temperature (1000°C - 1200°C) and strain rate (1 s^{-1} - 30 s^{-1}) are typical of hot forging operations of the three materials.

In the two-step compression experiments (Figure 3b), the specimen undergoes two deformation stages at the same temperature, but with different amounts of strain and strain rate in the two steps. Between the first and the second step of deformation the specimen is held at the testing temperature for 30 s. In all tests the total amount of strain equals 1 and the strain rate of the first step is assumed always equal to 10 s^1. The first part of the test represents the effects of a previous thermal and mechanical history.

The plan of all the experiments is summarised in Table 4.

For each of the testing conditions of Table 4, the microstructure of the material just before the deformation step has been analysed. To this aim, the specimen has been quenched in water from the testing temperature at the end of the soaking time (Figures 3a and 3b): The quenching has provided for the analysis of the material microstructure, allowing the determination of the average austenite grain size according to ASTM E112-88.

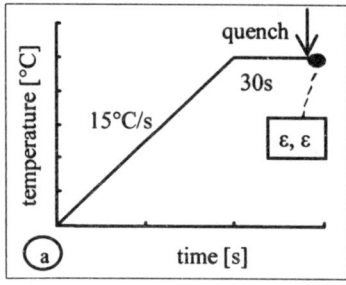 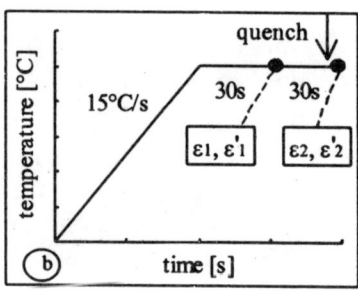

Figure 3. *Temperature-vs.-time diagram for a) single and b) two-steps experiments*

Table 4. *The experimental plan*

Materials: C70S; AISI 304;										
Single-step experiments				Two-steps experiments						
ε	ε' [s^{-1}]	Temperature [°C]			ε	ε' [s^{-1}]	ε	ε' [s^{-1}]	Temperature [°C]		
1	1	1000	1100	1200	0.3	10	0.7	1	1000	1100	1200
1	10	1000	1100	1200	0.3	10	0.7	10	1000	1100	1200
1	30	1000	1100	1200	0.3	10	0.7	30	1000	1100	1200

4. Results

Figures 4 and 6 show two flow curves: the upper one is relevant to a single-step experiment, while the lower curve to the second step of a two-step experiment. Both the curves are obtained for the values of temperature and strain rate. From the comparison of the two curves it can be appreciated the influence of a previous thermal and mechanical history on the flow strength: the material reacts differently when tested under the same current deformation conditions, but with different thermo-mechanical cycles; this difference is close to 25% for the stainless steel and 20% for the Nickel-based super alloy.

The previous cycles influence not only the mean value of the flow stress, but also evident is a shift in the value of the peak strain that implies a delay in the start of the phenomenon of the dynamic recrystallisation; this shift is close to 0.1 compared to the single-step peak strain for the superalloy. The differences in the values of flow stress and peak strain are essentially due to the different thermal and mechanical history.

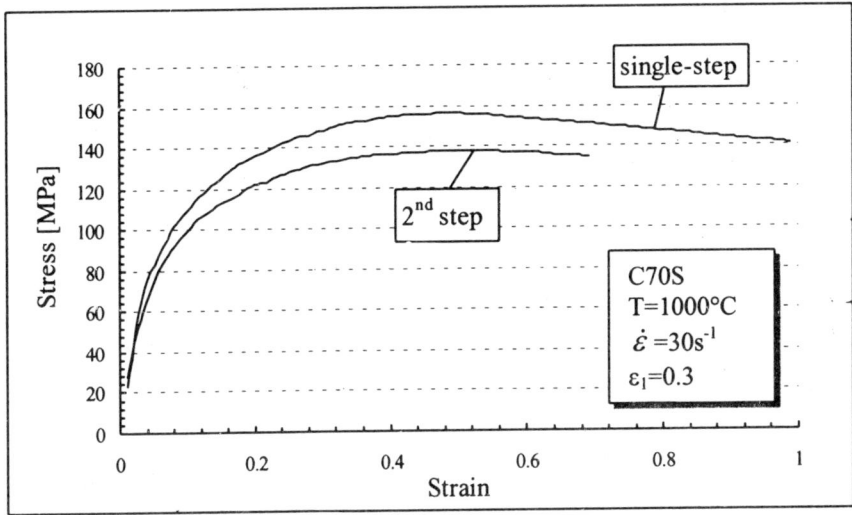

Figure 4. *Flow strengths in single- and two-steps experiments for C70S*

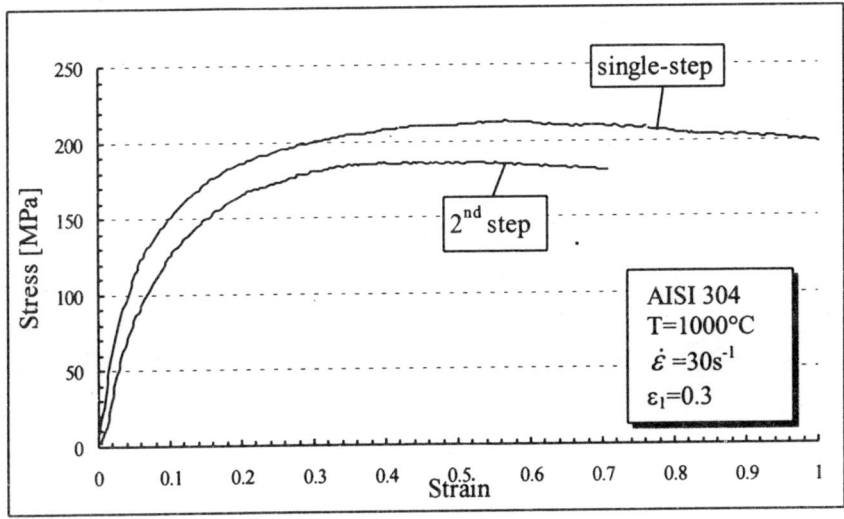

Figure 5. *Flow strengths in single- and two-steps experiments for AISI 304*

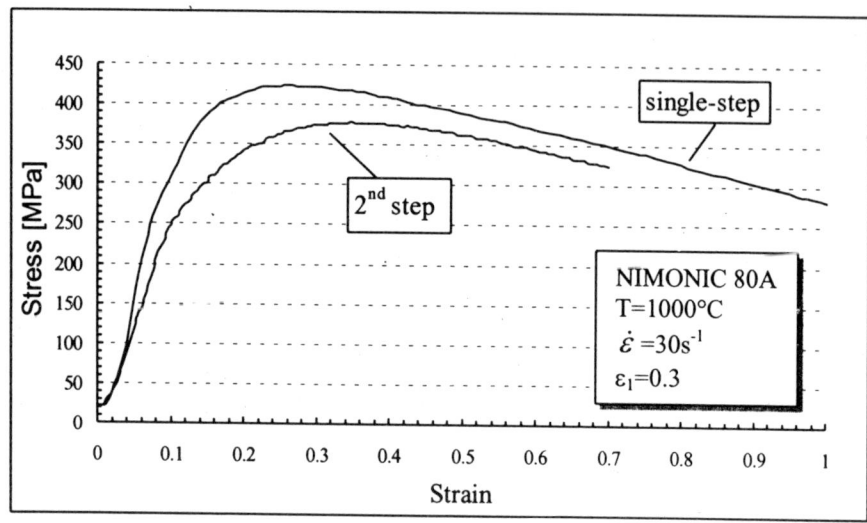

Figure 6. *Flow strengths in single- and two-steps experiments for Nimonic 80A*

In Figures 7 to 9 the microstructures of the three materials, observed in the samples quenched just before the single-step experiment and just before the second step of the two-steps experiment performed with the same process parameters, are compared. The difference between the average grain sizes is evident; that means the material microstructure is greatly influenced by the previous thermal and mechanical cycles.

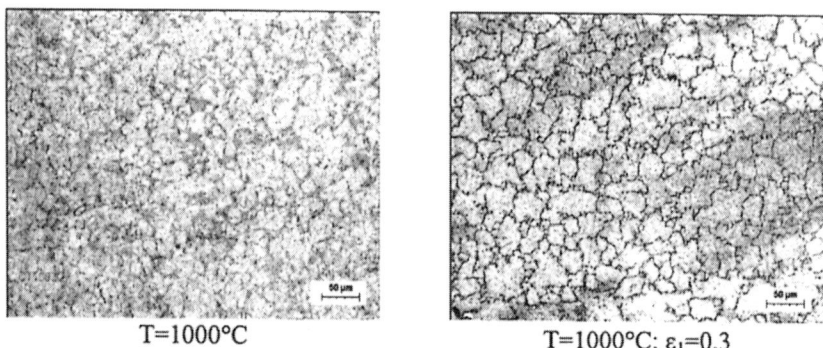

Figure 7. *Microstructures in single- and two-step experiments for C70S*

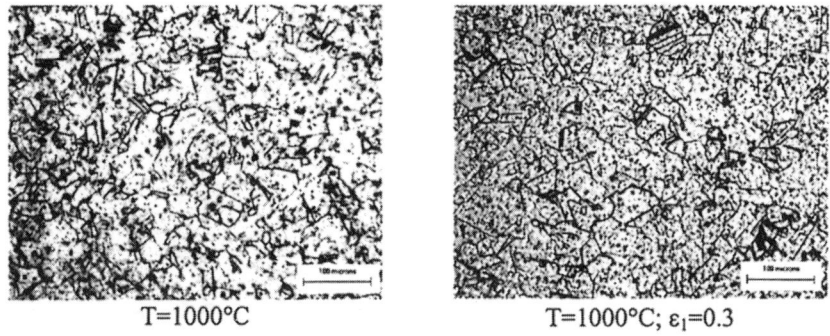

| T=1000°C | T=1000°C; ε_1=0.3 |

Figure 8. *Microstructures in single- and two-step experiments for AISI 304*

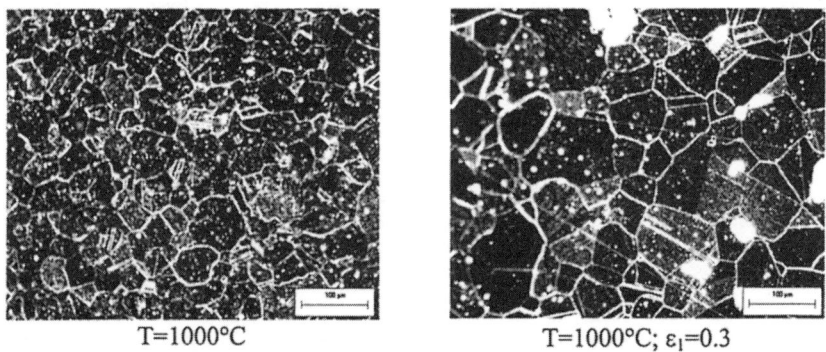

| T=1000°C | T=1000°C; ε_1=0.3 |

Figure 9. *Microstructures in single- and two-step experiments for Nimonic 80A*

These results are in agreement with the empirical equation:

$$\varepsilon_p = A \cdot D_0^{\,n} \cdot Z^m$$

that relates the peak strain with process parameters (Z) and initial microstructure (average grain size D); A, n and m are constants of the material.

Accordingly the microstructural parameter average grain size can suitably represent the first part of the cycle in the two-step experiment and therefore can summarise the effects of the previous history.

5. Concluding remarks

The two-step experiments demonstrated that the previous thermal and mechanical history affects value and location of the peak in the flow stress curve.

The testing procedures currently used to obtain flow curves prove to be more suitable to evaluating the sensitivity of the material flow strength to temperature and strain rate rather than to provide accurate description of rheological behaviour.

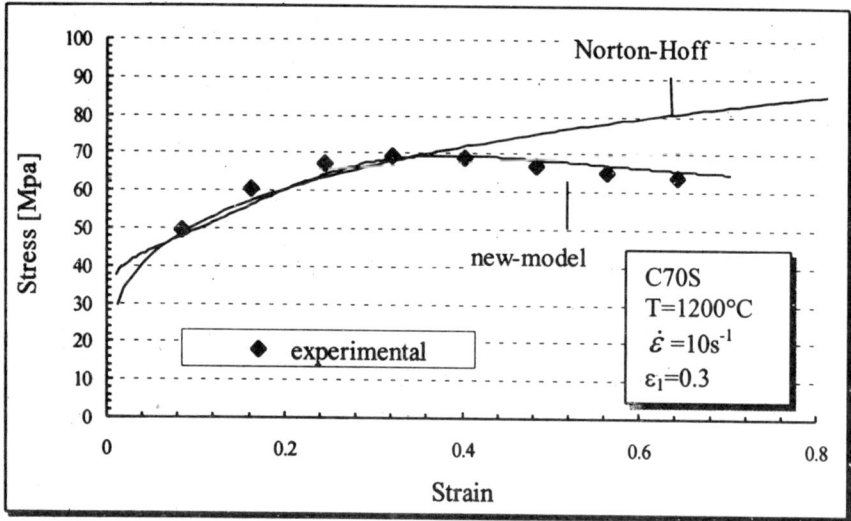

Figure 10. *Comparison between the flow stress evaluated through the Norton-Hoff equation, model [1] and the experimental results for C70S*

The equations that are currently used in FE codes are not suitable for describing accurately the rheological behaviour of hot forged materials. They do not take into account the effects of the softening and the previous history in hot forging conditions.

To overcome the above limitations the authors proposed recently [BAR 2000] the following material constitutive model:

$$\sigma = G \cdot F \cdot N_H \qquad [1]$$

where N_H is the Norton-Hoff law that relates the material flow stress to the instantaneous values of strain, strain rate and temperature; F is the function that permits to correctly locate the flow curve peak accommodating the effects of the softening phenomena; G is the function that takes into account the effects of the previous thermal and mechanical history through the microstructural parameter average grain size.

In Figure 10 the flow curve measured during the second step of a two-step experiment is compared with the one calculated using the Norton-Hoff model and model [1]. The proposed model is capable of accommodating the effects of the flow softening on the flow curve as well as the effects of a previous thermal and mechanical on the material flow stress.

6. References

[ALT 79] ALTAN.T., LAHOTI G.D., "Limitations, applicability and usefulness of different Methods in analysing forming problems", *Annals of CIRP vol. 28/2/1979*.

[BAR 98] BARIANI P.F, DAL NEGRO T., "Hot Deformation Studies on a Micro-Alloyed Steel: the Influence on the Flow Stress of the Varying Process Conditions", *Proc. 1st Esaform Conference, 1998*.

[BAR 99] BARIANI P. F., BRUSCHI S., DAL NEGRO T., "The need for new testing procedures in multi-step hot deformation studies", *Proc. ICTP 1999*.

[BAR 2000] BARIANI P.F., BRUSCHI S., DAL NEGRO T., "A new constitutive equation for steels in hot forging operations", *submitted Proc. CIRP 2000*.

[MAR 80] MARCINIAK Z., KONIECZNY A., "Analysis of Multi-Stage Deformation within the Warm-Forming Temperature Range", *Annals of CIRP Vol.29/1/1980*.

[YOS 94] YOSHINO M., SHIRAKASHI T., "Flow stress equation including effects of strain rate and temperature", *JSTP, 1994*.

[OH 95] OH S. I., "FEM Simulation of Hot Forging with Dynamic Recovery Process", *Annals of CIRP, 1995*.

[RAO 96] RAO K.P., PRASAD Y.K.D.V., "Hot deformation Studies on a Low-Carbon Steel: Flow Curves and the Constitutive relationship", *J. of Mat. Proces. Tech 56, 1996*.

[SEL 90] SELLARS C.M., "Modelling Microstructural Development during hot rolling", *Mater. Sci. Tech. 1990*.

[YOS 95] YOSHINO M., SHIRAKASHI T., "New Flow stress Equation of Ti-6Al-4V alloy", *J. of Mat. Proces. Tech. 48, 1995*.

Chapter 11

Measurement of Flow Stress and Critical Damage Value in Cold Forging

Victor Vazquez and Taylan Altan
Engineering Research Center for Net Shape Manufacturing (ERC/NSM), The Ohio State University, USA

1. Introduction

Flow stress is generally measured using a compression test instead of a tensile test because higher strains can be reached. Unfortunately, there is no standard flow stress testing protocol and due to differences in training of technicians different laboratories may report different results for the same materials. Additionally, the flow stress of the material is affected by any previous heat and/or mechanical treatments (like hot rolling, cold drawing, annealing, etc), so that it is not recommended to use the flow stress data of a material purely based on chemical composition. Accurate flow stress allows the designer to predict flow induced defects like folds or suck in defects.'

Due to strain hardening, in cold forging the formability of the material is more limited than in hot and warm forging (see Figure 1). A measure of the formability of a material is known as the *critical damage value* (CDV), which can be considered as a material constant, similar to yield stress or the tensile strength [CER 97].

There are several criteria to determine the CDV. Kim [KIM 94] developed a methodology to determine the CDV for several ductile fracture criteria (see Figure 2). It was found that Cockroft and Latham's criterion predicts with good agreement the amount of deformation and the location at which the fracture occurs by obtaining the maximum damage value (MDV) from FEM Simulations. If the MDV for the selected criterion coincides with the location of the cracks this is known as the CDV of the material.

In Cockroft and Latham's criterion fracture occurs when the cumulative energy due to the maximum tensile stress, σ^*, exceeds a certain value (equation 1). [OH 79] modified the criterion into a non-dimensional form (equation 2).

$$\int_{}^{\overline{\varepsilon}_f} \sigma^* d\overline{\varepsilon} = C_a \qquad [1]$$

$$\int_{}^{\overline{\varepsilon}} \frac{\sigma^*}{\overline{\sigma}} d\overline{\varepsilon} = C_b \qquad [2]$$

2. Measurement of flow stress

In a simple form, the flow stress ($\overline{\sigma}$) is concluded to be a function of strain ($\overline{\varepsilon}$), strain rate ($\dot{\overline{\varepsilon}}$), temperature (T), and the history or structure of the material (S), i.e.,

$$\overline{\sigma} = f(\overline{\varepsilon}, \dot{\overline{\varepsilon}}, T, S) \qquad [3]$$

In the range of practical strain rates, the flow stress is determined by conducting compression tests in a cam plastomer (constant $\dot{\varepsilon}$), a high speed hydraulic testing machine (constant $\dot{\varepsilon}$), or a mechanical press (varying $\dot{\varepsilon}$).

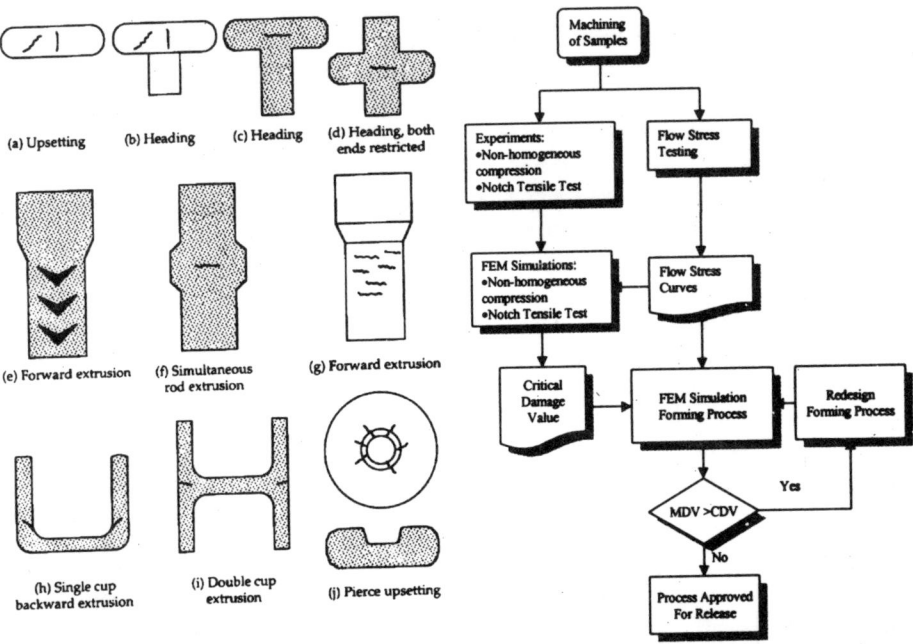

(a) Upsetting   (b) Heading   (c) Heading   (d) Heading, both ends restricted

(e) Forward extrusion   (f) Simultaneous rod extrusion   (g) Forward extrusion

(h) Single cup backward extrusion   (i) Double cup extrusion   (j) Pierce upsetting

Figure 1. *Typical cracks in cold forging operations (internal cracks are identified by shaded sections)*

Figure 2. *Methodology to predict and prevent the formation of cracks [KIM 94]*

Two different types of cylindrical specimens were used to determine the flow stress in the compression test: a) specimen with spiral grooves of about 0.010 inches deep (Figure 3a), and b) Rastegaev's specimen (Figure 3b).

To reduce the barreling effect during compression various lubricants are used, such as: a) paraffin, b) PTFE (sprayed), c) PTFE foils between dies and specimen.

3. Technique for measurement of Critical Damage Value (CDV)

The critical damage value is obtained using two tests; the non-uniform compression test with grooved dies and the tensile test with a notched specimen, see Figure 4. The non-uniform compression test with grooved dies consists of upsetting a cylinder until a crack initiates near the equatorial surface of the workpiece. The specimen height at which the crack occurs is known as the *fracture height*, H_f. Then FE simulations are conducted for the compression test up to the fracture height, H_f, to obtain the distribution of the damage value in the workpiece at the time of fracture. In the tensile test with notched specimens, the neck diameter is measured continuously until the occurrence of fracture (d_f). Then simulations are conducted to calculate the damage distribution at the instant the neck reaches the fracture diameter d_f.

4. Experimental study

Two different materials were selected in this study: SAE 1524 (spheroidised) which is a high-manganese carbon steel and SAE 1137 (hot rolled), which is a resulfurised carbon steel. The SAE 1137 material was produced as a coarse grain product, making it more likely to crack. The chemical compositions of these steels are given in Table 1.

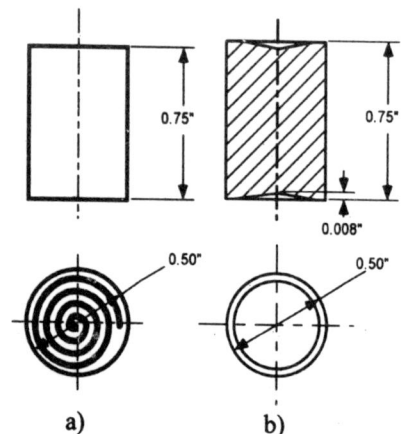

Figure 3. *a) spiral and b) Rastegaev's compression test specimens*

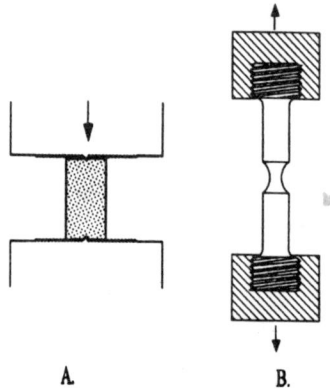

Figure 4. *Processes in order to determine critical damage value: A. Non-homogeneous compression test, B. Notch tensile test*

4.1 Experimental measurement of the flow stress by means of compression tests

The tests were conducted at room temperature using a 160-ton hydraulic press. At this temperature and pressing speed the effect of strain rate is negligible. Two types of specimens (Rastagaev's and spiral grooved) were used for the measurement of flow stress. Rastegaev's specimens performed better than spiral grooved specimens (see Figure 5, minor barreling and no skewing were observed)

The spiral grooved specimens were compressed to lower strains (~0.6 to 0.7) while the Rastegaev's specimens were compressed to larger higher strains.

Table 1. Chemical composition SAE 1524 (spheroidised) and SAE 1137 (hot rolled)

SAE	C	Mn	P	S	Si	Cu	Sn	Ni	Mo	Cr	As	Cb	V	N	Al
1524	.22	1.35	.040	.050	.15	---	---	---	---	---	---	.02	---	---	.015
1137	.39	1.48	.013	.10	.16	.17	.01	.06	.02	.08	.005	.001	.01	.004	.004

The flow stress data for SAE 1137 and SAE 1524 are shown in Figure 6. SAE 1137 shows a higher flow stress than SAE 1524 and also a greater strain hardening effect. The flow stress data shows good repeatability up to strains of 0.7. However, more variability is observed at larger strains; this is mainly due to deflection of the press and heating effects. Also the flow stress measured with the Rastegaev's specimens is slightly higher than that measured with the spiral specimens, approximately 4.5% for the SAE 1137 and 5.7% for SAE 1524. This error could be due to compensation of elastic deflection and measurement error (see Figure 6).

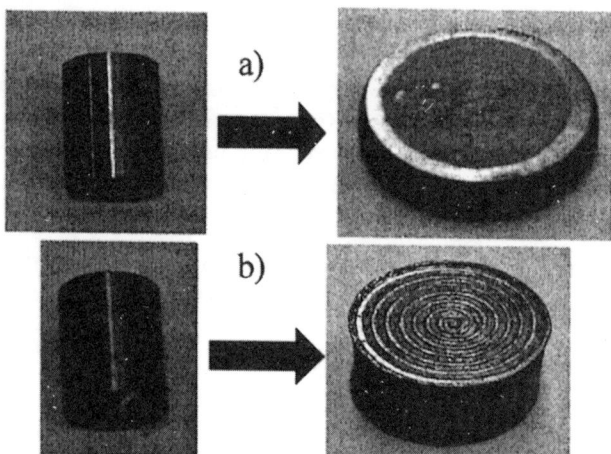

Figure 5. *a) Rastegaev's and b) spiral grooved specimens of material SAE 1137, hot rolled before and after deformation [SAN 99]*

4.2. Experimental measurement of critical damage value

4.2.1. Non homogeneous compression of cylinders

During non-homogeneous compression tests, the samples were inspected for cracks at 1-mm intervals of deformation. The results from these show that (see Figure 7): a) vertical cracks develop at the equatorial surface of the specimens, b) SAE 1524 can be reduced in height an average of 25% more than SAE 1137, c) at the moment of crack bursting, SAE 1524 showed a more sudden failure than SAE 1137.

4.2.2 Notch tensile test

The fracture found in the tension specimens is cup-and-cone type. This means that the fracture begins at the center of the notched specimen. The elongation of the specimen and the reduction of the neck diameter are higher for SAE 1524, indicating higher ductility.

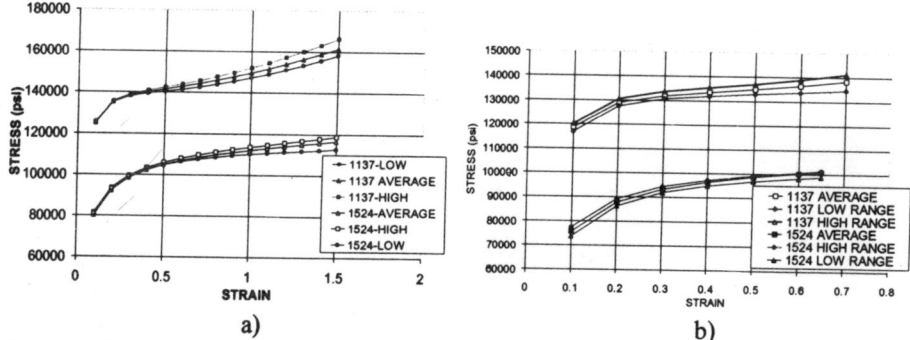

Figure 6. *Flow stress measurements SAE 1524 (annealed) and SAE 1137 (hot rolled): a) Rastegaev's specimens; b) Spiral grooved specimens [SAN 99]*

4.3. Calibration of critical damage value through process simulations

The calibration of the CDV is achieved through FEM simulations of the compression and tensile test. The distribution of the damage value can be seen in Figure 8. In the compression with grooved dies the MDV is at the equatorial surface, while in the notch tensile test it is at the center of the notched specimen.

The critical damage values for notched tensile test in the simulation are higher than the ones obtained from the compression with grooved dies. An explanation for this is that since the fracture in the tensile test originates from the center it is not possible to detect it at the time of initiation. In order to overcome this problem the cracking feature of DEFORM™ 2D and a CDV equal to that of the compression test were used to simulate the tensile test. This damage value is modified until the neck diameter in the simulation just before cracking matches that measured in the experiment. The CDV's obtained for both tests through this procedure are given in Table 2. Figure 9 shows an example of crack initiation and propagation at the neck of specimen during testing..

5. Conclusions

– Two different types of specimens were used to determine the flow stress of two different materials: SAE 1137 (hot rolled) and SAE 1524 (spheroidised annealed).

– Critical damage values (CDV) were determined by conducting non-uniform compression tests with grooved dies and notched tensile tests, and these results were calibrated by means of FEM simulations.

- Additional refinement of the physical testing procedure for CDV will improve the accuracy of predicting defects. These refinements could be better, by developing a more accurate detection of crack initiation.

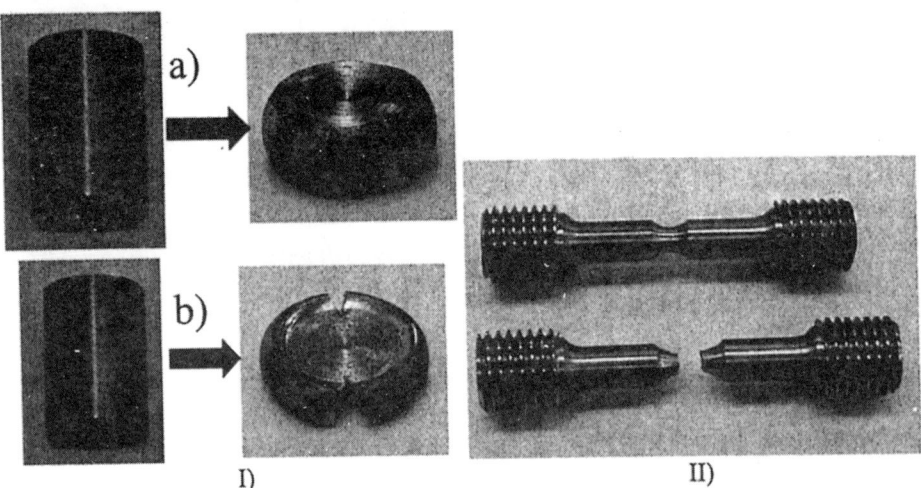

I)   II)

Figure 7. *I) Specimens for non-uniform compression test, a) SAE 1137 hot rolled, b) SAE 1524 spheroidised annealed, II) Notched tensile test specimens [SAN 99]*

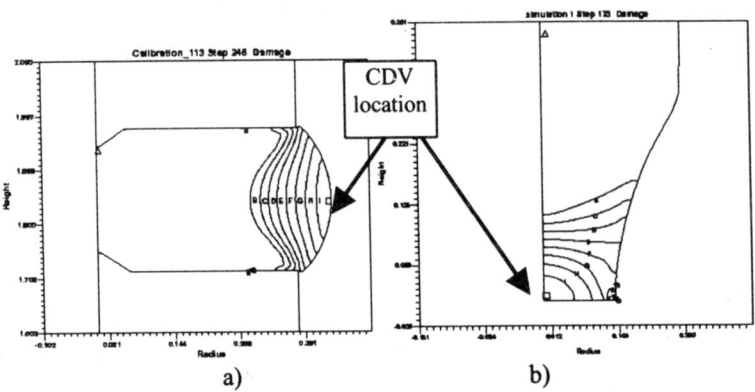

a)   b)

Figure 8. *Distribution of the critical damage value for material SAE 1137 (hot rolled) a) non-homogeneous compression and b) notched tensile test [SAN 99]*

Table 2. *Critical damage values obtained during the FEM calibration [SAN 99]*

Material	Method	CDV	Location
SAE 1137	compression with grooved dies	.37	equatorial surface
hot rolled	notch tensile test	.35	center of specimen
SAE 1524	compression with grooved dies	.65	equatorial surface
Spheroidised	notched tensile test	.675	center of specimen

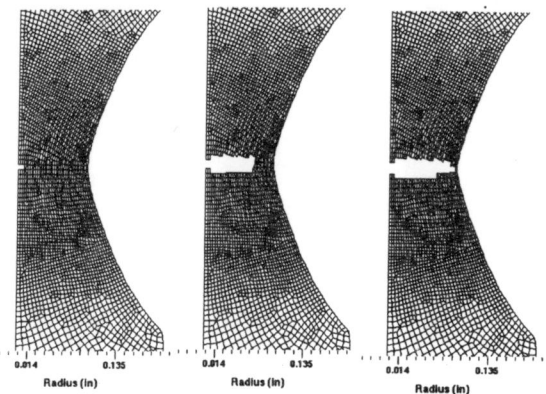

Figure 9. *Fracture behaviour for tensile test for SAE 1524, CDV=0.675 [SAN 99]*

6. References

[ALT 73] ALTAN T., BOULGER F.W., *"Flow stress of metals and its application in metal forming analyses"*, Contributed by the Production Engineering Division and presented at the Production Engineering Conference, May 16, 1973, The American Society of Mechanical Engineering, Paper No. 73-Prod-4.

[KIM 94] KIM H., YAMANAKA M., ALTAN T., *"Prediction of ductile fracture in cold forging by FE simulations."* Engineering Research Center for the Net Shape Manufacturing. Report No. ERC/NSM-94-42. August 1994.

[KIM 94] KIM H., YAMANAKA M., ALTAN T, ISHII K., *"Effects of design and process parameters upon the formation of chevron defects in forward rod extrusion"*, Engineering Research Center for the Net Shape Manufacturing. Report No. ERC/NSM-94-41, August 1994.

[CER 97] CERETTI E., TAUPIN E., ALTAN T., *"Simulation of metal flow and fracture applications in orthogonal cutting, blanking and cold extrusion"*, Annals of the CIRP, Vol. 46/1/1997.

[OH 79]   OH S. I., CHEN C., and KOBAYASHI S., Ductile fracture in Axisymmetric Extrusion and Drawing. Part 2 Workability in Extrusion and Drawing, Transactions of the ASME, 101, Feb.,1979, pp.36-44.

[SAN 99] SANTIAGO-VEGA C., VAZQUEZ V., ALTAN T., Development of Design Guidelines to Avoid Chevron Crack Formation in Forward Extrusion with Spherical Dies, Report No. DaimlerChrysler-F/ERC/NSM-99-R-27.

Index

168   Friction and Flow Stress in Forming and Cutting